Surveying Review
for the
Civil Engineer

Jack Liu

Second Edition

ISBN 1-57645-017-1

Engineering Press   Austin, Texas

ISBN 1-57645-017-1

Engineering Press
P.O. Box 200129
Austin, TX 78720-0129

PREFACE

Effective January 1, 1988, in California, all civil engineer examinees are required to sit for Special Civil Engineer (CE) Exam which tests the applicants' knowledge of seismic principles and engineering surveying principles. The Special CE Exam is a supplement to the 8-hour National Council of Engineering Examiners (NCEE) examination.

Since the Special Civil Engineer Exam has been required, passing rates for the Civil Engineer Exam have fallen dramatically (7 % in April, 1988 and 8 % in October, 1989).

In October, 1990, a new test plan was made by the Board of Registration for Professional Engineers and Land Surveyors. The Special Civil Engineer Exam is a four-hour examination consisting of two sections, Seismic Principles and Engineering Surveying. Examinees will be allotted two hours to complete each exam section which includes 20-40 multiple-choice questions and 1-3 design problems. Each section is open book. In October, 1991, the engineering surveying principles section is composed entirely of multiple-choice questions.

The Registration Board requires applicants to demonstrate the knowledge to carry out the following responsibilities:

1. RESPONSIBLLITIES AND TASKS COMMON TO   42%
   ENGINEERING SURVEYING

2. RESPONSIBILITIES AND TASKS PERTAINING TO   38%
   TOPOGRAPHIC SURVEYING

3. RESPONSIBILITIES AND TASKS PERTAINING TO   19%
   CONSTRUCTION SURVEYING

Examinees should bring the following materials to the exam administration in addition to reference materials:

   No. 2 pencil, protractor, triangle, engineer's scale.

Although the survey exam is open-book, there is no time to read the Surveying Manual during the examination hours. The examination is very fast-paced. Solving the problems is a good method for preparing the exam. In order to fully understand the examples in this book, you should try to solve the problems by yourself instead of reading the solution first.

TABLE of CONTENTS

1. HORIZONTAL CURVE

Symbols

PI Point of Intersection.
I   Intersection angle at PI (= Central angle).
T   Tangent distance from BC to PI (or from EC to PI).
E   External distance, distance from PI to midpoint of curve.
M   Middle ordinate, distance from midpoint of curve to
    midpoint of long chord.
L   Length of curve, distance from BC to EC along the arc.
BC Beginning of Curve. (also termed PC)
EC End of Curve. (also termed PT)
O   Radius Point (center of radius).
LC Length of the Long Chord.
R   Radius of the curve

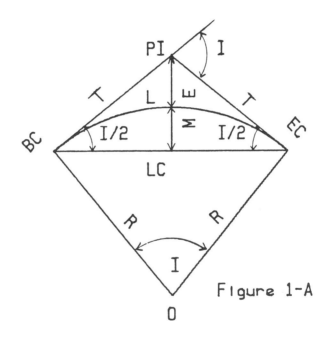

Figure 1-A

$$T = R \tan (I/2)$$

$$L = R\,I \quad (I \text{ in radians})$$

$$LC = 2 R \sin (I/2)$$

$$\frac{R}{R + E} = \cos \frac{I}{2} \quad ; \quad E = R \left(\sec \frac{I}{2} - 1\right)$$

$$\frac{R - M}{R} = \cos \frac{I}{2} \quad ; \quad M = R \left(1 - \cos \frac{I}{2}\right) \quad \text{(Eqs.1-1)}$$

There are two types of horizontal curves: Circular Arcs and Spirals. A circular arc connecting two tangents is a simple curve. Easement curves are used to lessen the sudden change in curvature at the junction of a tangent and a circular curve. A spiral makes a good easement curve.

There are two definitions of 'Degree of Curve':
The CHORD definition is used in railroad practice.
The ARC definition is used for highway work.

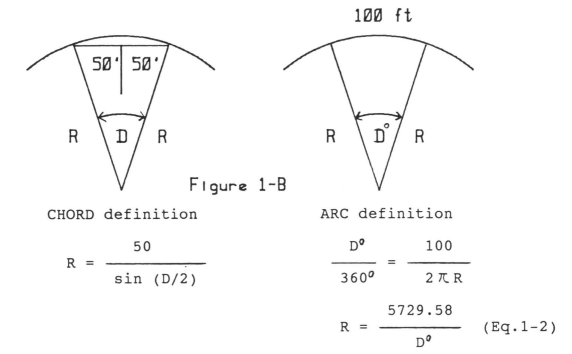

Figure 1-B

CHORD definition

$$R = \frac{50}{\sin (D/2)}$$

ARC definition

$$\frac{D^o}{360^o} = \frac{100}{2 \pi R}$$

$$R = \frac{5729.58}{D^o} \quad (Eq.1-2)$$

The DEFLECTION ANGLE formed by a tangent and chord is equal to one-half the intercepted arc. This is illustrated in the following figure:

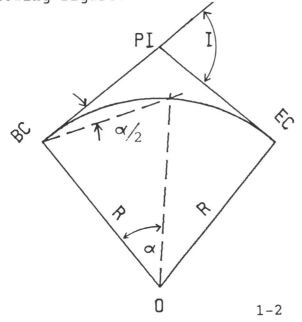

1-2

Example 1-1
_____

Given a curve with a radius R = 350 ft, calculate the
deflections to each full station and the E.C.(End of Curve).
The B.C.(Beginning of Curve) is station 12+13.22

Station	Deflection
B.C. 12+13.22	
13+00	-----------
14+00	-----------
15+00	-----------
E.C. 15+19.96	-----------

Solutions:
=============

a. Arc length = L $\Rightarrow$ $\dfrac{D^o}{360^o} = \dfrac{L}{2 \pi R}$

$$D^o = \dfrac{360^o L}{2 \pi R} \qquad\qquad R = 350' \rightarrow D^o$$

b. Deflection = $\dfrac{D^o}{2} = \dfrac{1}{2} \left(\dfrac{360^o L}{2 \pi R} \right) = \dfrac{90^o L}{\pi R} = \dfrac{90^o L}{350 \pi}$

The deflections to each station:

Station	L (ft)	Deflection
B.C. 12+13.22	0.00	$0^o 00' 00''$
13+00	86.78	$7^o 06' 11''$
14+00	186.78	$15^o 17' 17''$
15+00	286.78	$23^o 28' 24''$
E.C. 15+19.96	306.74	$25^o 06' 25''$

Example 1-2

An existing 5⁰ curve connects B.C. and E.C. as shown in Fig.1-C.
In order to join a proposed highway, the tangent of the curve
is relocated. The new tangent line is 150 feet away, and is
parallel to the existing tangent as shown in Figure 1-C. Find

  (A) the existing B.C., E.C. station.

  (B) the radius of curvature for the new curve.

  (C) the E.C., middle ordinate, and external distance
      of the new curve.

  (D) the station at the intersection of the existing curve
      and the new tangent.

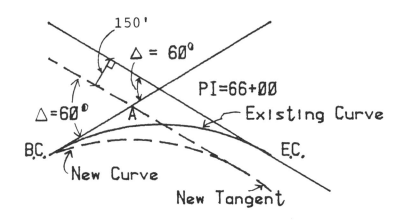

Figure 1-C

Solutions:
===

(A) *Existing curve, from Eq.(1-2),*

$$R = \frac{5729.58}{D^0} = \frac{5729.58}{5^0} = 1145.916'$$

$T = R \tan (I/2) = (1145.916) \tan(60^0/2) = 661.6'$

$L = \triangle (R) = 2\pi(60^0/360^0)(1145.916) = 1200'$

Station:

$B.C. = P.I. - T = (66+00) - (6+61.6) =$ $\boxed{59+38.4}$

$E.C. = B.C. + L = (59+38.4) + (12+00) =$ $\boxed{71+38.4}$

1-4

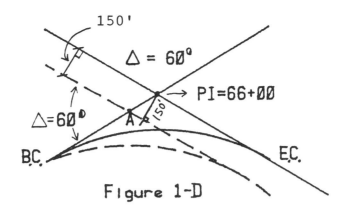

Figure 1-D

(B) New curve

From Fig.1-D,

$\sin 60^{\circ} = 150'/(A \text{ to } P.I.)$

A to P.I. = 173.205'

New Tangent T_n = T - (A to P.I.)

T_n = 661.6 - 173.205 = 488.4'

New Radius $R_n = \dfrac{T_n}{\tan (I/2)} = \dfrac{488.4}{\tan 30^{\circ}} = \boxed{845.93 \text{ ft}}$

(C) New curve

$L_n = \triangle (R_n) = 2\pi(60^{\circ}/360^{\circ})(845.93) = 885.856'$

New E.C. = B.C. + L_n = (59+38.4) + (8+85.86)

$= \boxed{68+24.26}$

Middle ordinate, $M = R_n(1 - \cos \dfrac{I}{2})$ $\qquad$ (Eqs.1-1)

$= 845.93 (1 - \cos 30^{\circ}) = \boxed{113.33 \text{ ft}}$

External distance, $E = R_n(\sec \dfrac{I}{2} - 1)$

$= 845.93 (\dfrac{1}{\cos 30^{\circ}} - 1) = \boxed{130.87 \text{ ft}}$

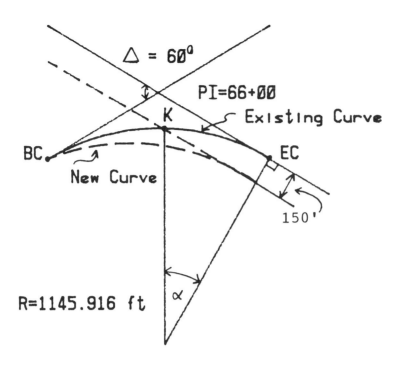

$\triangle = 60^0$

PI=66+00

K

Existing Curve

BC

New Curve

EC

150'

R=1145.916 ft α

(D) Station K (at the intersection of the existing curve
and the new tangent)

$$\cos \alpha = \frac{R - 150}{R} = \frac{1145.916 - 150}{1145.916} \implies \alpha = 29.646^0$$

Curve length from K to E.C.

$L_K = \alpha (R) = 2 \pi (29.646^0/360^0)(1145.916) = 592.91'$

Station K = E.C. - L_K = (71+38.4) - (5+92.91)

$= \boxed{65+45.5}$

Example 1-3
―――――――

A highway center line begins at point 'A' (B.C.) which is due
south of the center of a 1000-foot radius curve, thence along
the curve concave to the northwest to point 'B' (P.R.C.),
thence on a curve concave to the south (radius 600 ft) to point
'C' (E.C.), station 104+00, thence on a tangent to point 'D',
station 109+00. The bearing of C.D. is S 75°E and point 'D'
is directly east of point 'A'.

REQUIREMENT
―――――――――――

It is proposed to relocate the highway center line to a new
location which is a straight line between points 'A' and 'D'.
what is the length AD of this new alignment?

Solutions:

$$\overline{QA} = \overline{QB} = 1000'\ (radius)$$

$$\overline{CF} = \overline{BF} = 600'\ (radius)$$

$$\overline{CD} = (109+00) - (104+00)$$

$$= 500'$$

$$\overline{QF} = \overline{QB} + \overline{BF} = 1600'$$

$$x = \overline{DO}$$

$$y = \overline{Of}$$

$$z = \overline{fA}$$

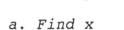

$$\overline{AD} = x + y + z$$

a. Find x

$$x = \overline{CD}/(cos\ 15^o) = 500\ /(cos\ 15^o) = 517.638'$$

b. Find y

$$\overline{CO} = \overline{CD}\ tan\ 15^o = 500(tan\ 15^o) = 133.975'$$

$$y = (\overline{CF} - \overline{CO})\ sin\ 15^o = (600 - 133.975)\ sin\ 15^o = 120.616'$$

c. Find z

$$\overline{QE}^2 + z^2 = \overline{QF}^2$$

$$\overline{QE} = \overline{QA} + \overline{AE}\ ;\qquad \overline{QF} = \overline{QB} + \overline{BF}$$

$$\overline{AE} = \overline{Ff} = y/(tan\ 15^o) = 120.616\ /(tan\ 15^o) = 450.145'$$

$$(\overline{QA} + \overline{AE})^2 + z^2 = (\overline{QB} + \overline{BF})^2$$

$$(1000 + 450.145)^2 + z^2 = 1600^2$$

$$z = 676.076'$$

d. $\overline{AD} = x + y + z = $ | 1314.33 ft |

Example 1-4

Line ABCD in the Figure below represents a center line of a survey for a certain part of a construction project. Curve BC has a radius of 500', with center at O. Another construction line EG, intersects line ABCD at F.

The elevation of B is 100 ft and the grade toward C is + 1.0 %. The elevation of E is 93 ft and the grade toward F is + 0.5 %.

All other data is shown on the figure.

REQUIRED

What is the difference in elevation between the two construction lines at point F?

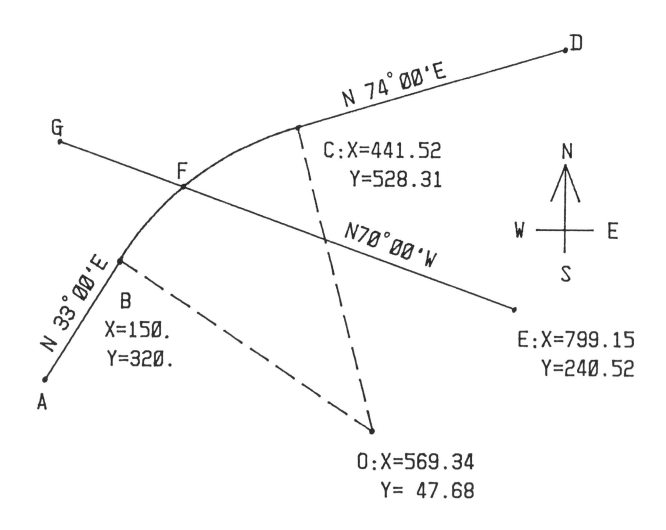

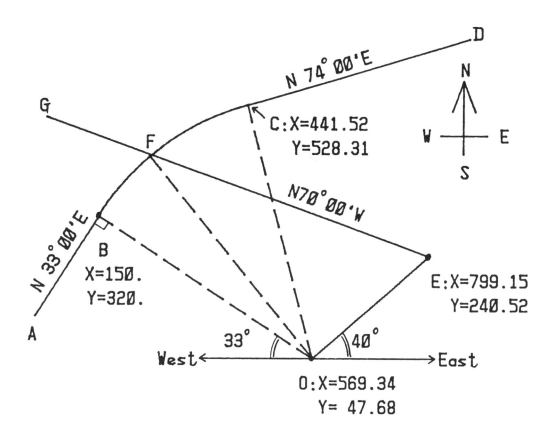

Figure 1-E

Solutions:

(A) *Calculate the elevation of F on line $\overline{BF}$:*

The grade from B toward F is 1%, length $\overline{BF}$ = 500 ($\angle$ BOF)

a. Find $\angle$ BOF

$\angle$ BOF = 180° – 33° – $\angle$ FOE – $\angle$ EON

b. Find $\angle$ EON

$$\angle \text{ EON} = \tan^{-1} \frac{240.52 - 47.68}{799.15 - 569.34} = \tan^{-1} \frac{192.84}{229.81} = 40°$$

c. Find $\angle$ FOE

$\angle$ FOE = 180° – $\angle$ EFO – $\angle$ FEO

$\angle$ FEO = $\angle$ EON + (90° – 70°) = 40° + 20° = 60°

1-9

d. Find $\angle EFO$ (See Fig.1-E)

In $\triangle$ FOE, $\dfrac{sin\ (\angle\ EFO)}{\overline{OE}} = \dfrac{sin\ (\angle\ FEO)}{\overline{FO}}$

$\overline{OE} = \sqrt{(229.81)^2 + (192.84)^2} = 300$; $\overline{FO} = 500$

Therefore, $\dfrac{sin\ (\angle\ EFO)}{300} = \dfrac{sin\ (60^0)}{500}$

$\angle EFO = 31.3064^0$

e. Calculate the elevation of F on line $\overline{BF}$:

From the above results,

$\angle FOE = 180^0 - \angle EFO - \angle FEO = 88.6936^0$

$\angle BOF = 180^0 - 33^0 - \angle FOE - \angle EON = 18.3064^0$

$\overline{BF} = 500\ (\angle\ BOF) = 500\ (18.3064^0)(2\pi)/360^0 = 159.75$ ft

The grade from B toward F is 1%

The elevation of F on line $\overline{BF}$:

$100 + 0.01(159.75) =$ $\boxed{101.598\ ft}$

(B) Calculate the elevation of F on line $\overline{EF}$:

In $\triangle$ FOE, $\dfrac{sin\ (\angle\ FOE)}{\overline{FE}} = \dfrac{sin\ (\angle\ FEO)}{\overline{FO}}$

$\overline{FO} = 500$; $\angle FOE = 88.6936^0$; $\angle FEO = 60^0$

Therefore, $\overline{FE} = 577.2$ ft

The grade from E toward F is 0.5%

The elevation of F on line $\overline{EF}$:

$93 + 0.005(577.2) =$ $\boxed{95.886\ ft}$

(C) the difference in elevation between line $\overline{BF}$ and line $\overline{EF}$ at point F is

$101.598 - 95.886 =$ $\boxed{5.712\ ft}$

Example 1-5

The centerline of an aqueduct was originally laid out as a reversed curve as indicated by the existing ₵ in the plan below. A construction project in the vicinity requires the aqueduct to be realigned. The new alignment will connect the existing tangents with a 1200 ft radius curve. This realignment will move the curve part of the ₵ back away from the proposed project and it will also replace the reversed curve that now exists.

REQUIREMENT

(A) What is the central angle of the new curve?

(B) Locate the B.C. of the new curve using the existing B.C. as the point of reference.

(C) Locate the E.C. of the new curve using the existing E.C. as the point of reference.

(D) What is the saving in distance between the new and the old alignments? Base your calculations on the distance from the B.C. of the new curve to the E.C. of the existing alignment.

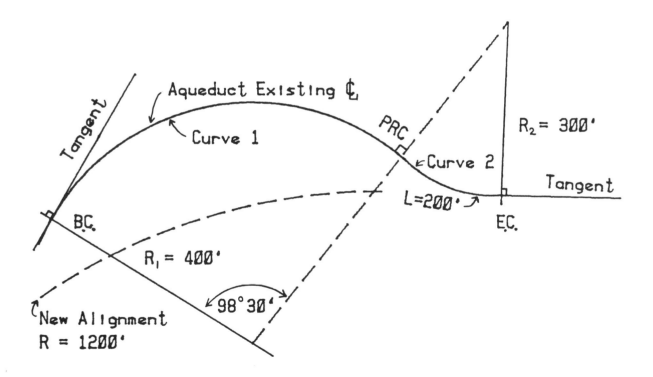

Existing B.C.: point G;   Existing E.C.: point F

$R_1 = \overline{GO} = 400'$; $R_2 = \overline{QF} = 300'$; $\overline{OQ} = 400' + 300' = 700'$

Q: Center of radius 300'

O: Center of radius 400'

$r = \angle QOG = 98°30'$

$L_2 = 200' = R_2(q)$

$q = \angle FQO$

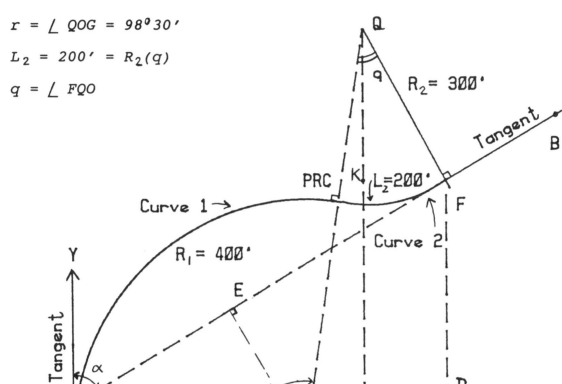

Figure 1-F

Solutions:

(A) Draw $\overline{OE} \parallel \overline{QF}$

α = The central angle of the new curve = $r - q$

$$q = \frac{L_2}{R_2}\left(\frac{360}{2\pi}\right) = \frac{200}{300}\left(\frac{360}{2\pi}\right) = 38°11.8'$$

$\alpha = r - q = 98°30' - 38°11.8' = \boxed{60°18.2'}$ (= 60.3°)

(B) New curve, $T = R \tan(\alpha/2) = 1200(\tan 30^{\circ}9.1') = 697.062'$

From the new B.C. to the existing B.C. $= T - \overline{AG}$

and $\overline{AG} = \overline{QP} - \overline{QK} - \overline{FD}$ (See Fig.1-F)

a. Find $\overline{QP}$ (See Fig.1-G)

$\overline{QP} = \overline{QO} \sin (\angle QOP)$; $\angle QOP = 180^{\circ} - r = 81^{\circ}30'$

$\overline{QP} = 700 \sin(81.5^{\circ}) = 692.311'$

$r = \angle QOG = 98^{\circ}30'$

$q = \angle FQO = 38^{\circ}11.8'$

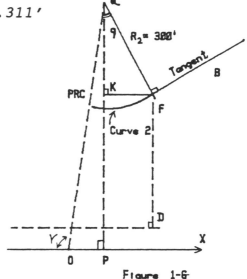

Figure 1-G

b. Find $\overline{QK}$ (See Fig.1-G)

$\overline{QK} = \overline{QF} \cos (\angle FQK)$; $\angle FQK = q - \angle OQK$

$\angle OQK = r - 90^{\circ} = 8^{\circ}30'$; $\angle FQK = q - 8^{\circ}30' = 29^{\circ}41.8'$

$\overline{QK} = 300(\cos 29^{\circ}41.8') = 260.598'$

c. Find $\overline{FD}$ (See Fig.1-F)

$\overline{FD} = \overline{AD} \tan (\angle BAD)$; $\angle BAD = 90^{\circ} - \alpha = 29^{\circ}41.8'$

$\overline{AD} = R_1 + \overline{OP} + \overline{KF}$

$\overline{OP} = \overline{QO} \cos(\angle QOP) = 700 \cos(81.5^{\circ}) = 103.467'$

$\overline{KF} = \overline{QF} \sin(\angle KQF) = 300 \sin(29^{\circ}41.8') = 148.622'$

Therefore, $\overline{AD} = 400 + 103.467 + 148.622 = 652.089'$

$\overline{FD} = 652.089 (\tan 29^{\circ}41.8') = 371.895'$

d. $\overline{AG} = \overline{QP} - \overline{QK} - \overline{FD} = 692.311 - 260.598 - 371.895 = 59.818'$

From the new B.C. to the existing B.C.

$= T - \overline{AG} = 697.062 - 59.818 =$ $\boxed{637.24 \text{ ft}}$

(C) From the new E.C. to the existing E.C. = $\overline{AF}$ - T   (See Fig.1-F)

$$\overline{AF} = \frac{\overline{AD}}{\cos (\angle BAD)} = \frac{652.089}{\cos (29^\circ 41.8')} = 750.684'$$

$\overline{AF}$ - T = 750.684 - 697.062 = $\boxed{53.622 \text{ ft}}$

(D) Distance saving:

(Base on the distance from the new B.C. to the existing E.C.)

a. New curve

L_3 = 1200(a) = 1200(2 π)(60.3°) /360° = 1262.92'

From the new B.C. to the existing E.C.

= (New curve L_3) + (New E.C. to existing E.C.)

= 1262.92 + 53.622

= 1316.54 ft

b. Existing curve

L_1 = 400(r) = 400(2 π)(98.5°) /360° = 687.66'

From the new B.C. to the existing E.C.

= (New B.C. to G) + (Existing curve L_1) + L_2

= 637.24 + 687.66 + 200

= 1524.9 ft

c. Saving in distance

1524.9 - 1316.54 = $\boxed{208.36 \text{ ft}}$

Example 1-6
─────────────

A loop ramp at an interchange was designed as a broken-back curve.
See Fig.1-H. The short tangent section is to be eliminated by
compounding the two given curves with a curve of 200' radius.

REQUIRED
─────────────

Calculate the stationing of the points of compoud curvature
(P.C.C.), and show the equation in stationing at the second
P.C.C.

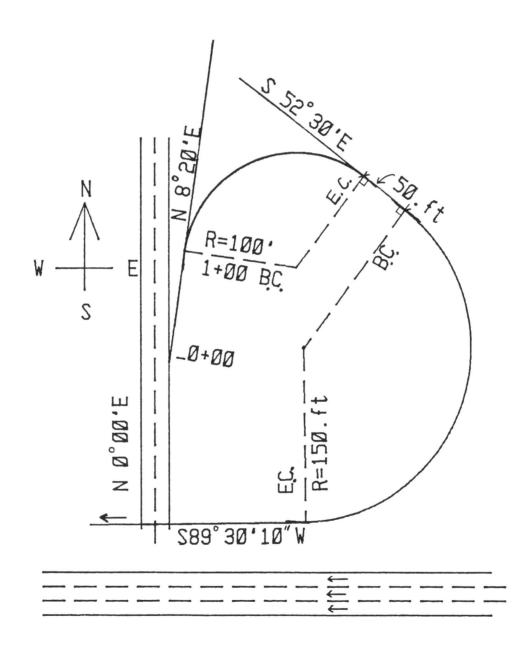

Figure 1-H

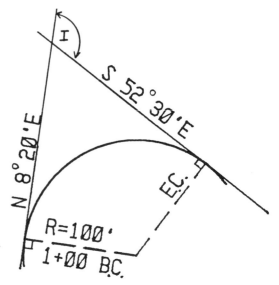

Solutions:

(A) Stationing along broken-back curve.

    a. *Curve of 100′ radius*

        Intersection angle I = 180⁰ – 8⁰20′ – 52⁰30′ = 119⁰10′

        $L = I\ (R) = 2\pi(119.167°/360°)(100) = 207.99′$

        E.C.1 = (1+00) + (2+7.99) = 3+7.99

    b. *Curve of 150′ radius*

        B.C.2 = (3+7.99) + (0+50) = 3+57.99

(B) Stationing the points of the compound curve:

    a. *Draw the following figure*

        In order to calculate the stationing of the P.C.C.,
        angles p and q are needed.

        p = 180⁰ – 90⁰ – 45⁰ – A

        q = 180⁰ – 45⁰ – B

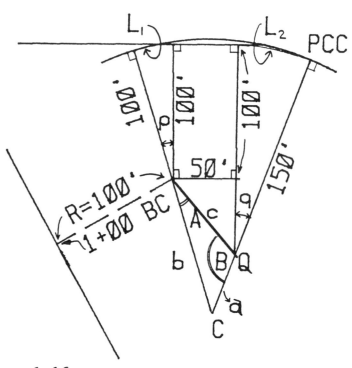

b. Find angles A and B

Given: $a = 50'$
$b = 100'$
$c = 50(\sqrt{2}) = 70.7106'$

Law of Cosines

$$\cos A = \frac{b^2 + c^2 - a^2}{2\,b\,c} = \frac{100^2 + 70.7106^2 - 50^2}{2\,(100)\,(70.7106)}$$

$\Rightarrow A = 27.8855^o = 27^o53'8"$

Law of Sines

$$\frac{a}{\sin A} = \frac{c}{\sin C} \Rightarrow \frac{50}{\sin 27.886^o} = \frac{70.7106}{\sin C}$$

$\Rightarrow C = 41.409^o = 41^o24'34"$

$B = 180^o - A - C = 110^o42'18"$

c. Find curve length L_1 , L_2

$p = 180^o - 90^o - 45^o - A = 17^o6'52"$

$q = 180^o - 45^o - B = 24^o17'42"$

$L_1 = p\,(100) = 2\,\pi\,(17.114^o/360^o)(100) = 29.87'$

$L_2 = q\,(150) = 2\,\pi\,(24.295^o/360^o)(150) = 63.604'$

d. Find station F, and G along broken-back curve

Station $F = E.C.1 - L_1 = (3+7.99) - (0+29.87)$

$= \boxed{2+78.12}$

Station $G = B.C.2 + L_2 = (3+57.99) + (0+63.6)$

$= \boxed{4+21.59}$

e. Find station G along the compound curve

$L_3 = C\,(200) = 2\,\pi\,(41.409^o/360^o)(200) = 144.55'$

Station G = Station $F + L_3 = (2+78.12) + (1+44.55)$

$= \boxed{4+22.67}$

Sample Problems 1

Problem A
_____

A horizontal curve is shown in the Figure below. PI is
located in the lake and inaccessible. Length of $\overline{AB}$ has
been found to be 614.75 ft.

REQUIRED
_____

   1. Determine the intersection angle at PI.

      (A) $40^0 51'$ (B) $22^0 37'$ (C) $18^0 14'$ (D) $4^0 23'$

   2. Find the radius of the horizontal curve (ft).

      (F) 2031.30   (G) 2030.30   (H) 2030.03   (J) 2033.31

   3. Determine the length of the horizontal curve (ft).

      (A) 1445.74   (B) 1474.54   (C) 1447.54   (D) 1454.74

   4. Find the PT station.

      (F) 75+89.75   (G) 79+58.74

      (H) 79+60.54   (J) 79+67.74

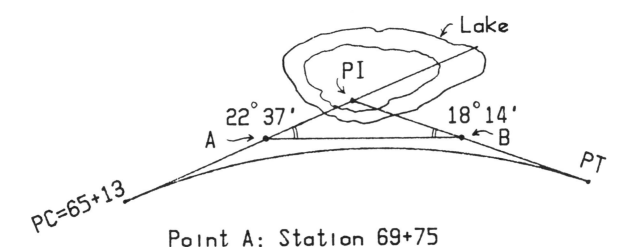

Point A: Station 69+75

Problem B

A highway right of way cuts across the west end of a property parcel. Half the width of the right of way is 50' as shown in the sketch below. The radius to the highway curve ℄ is 400' and the total central angle is 30°. The lot lines along the north and the south boundaries are parallel, are 100' apart. and lie on a line due East - West. The east property line lies on a bearing of N 30° E as shown. The intersection of the east right of way line and the south property line becomes the PC of a circular curve which will become the new western lot boundary.

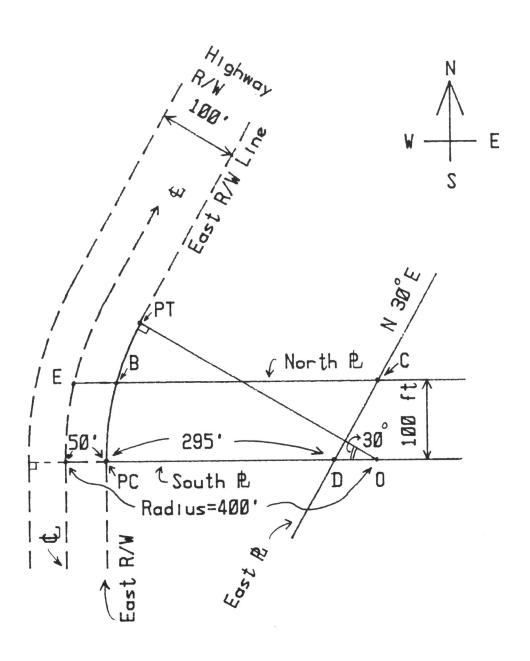

5. What is the central angle PC - O - B?

    (A) 15.0⁰   (B) 15.6⁰   (C) 16.0⁰   (D) 16.6⁰

6. What is the arc length PC - B which becomes the new western property line?

    (F) 101.0'   (G) 101.4'   (H) 103.5'   (J) 105.1'

7. What is the length of the north property line of the shortened lot? ($\overline{BC}$ =?)

    (A) 336.8'   (B) 335.2'   (C) 338.2'   (D) 337.0'

8. What is the length of the east property line? ($\overline{CD}$=?)

    (F) 115.5'   (G) 114.5'   (H) 111.5'   (J) 116.5'

9. What is distance measured along the north property line from the highway ₵ to the east R/W line? ($\overline{EB}$ =?)

    (A) 50.2'   (B) 51.0'   (C) 51.9'   (D) 53.0'

Problem C

In the figure below, the angles indicated thus ⌐ are 90⁰.

10. Calculate the length of the line marked b (ft).

    (A) 137.02   (B) 137.12   (C) 137.36   (D) 136.84

11. Calculate the length of the line marked a (ft).

    (F) 146.80   (G) 146.98   (H) 146.33   (J) 146.19

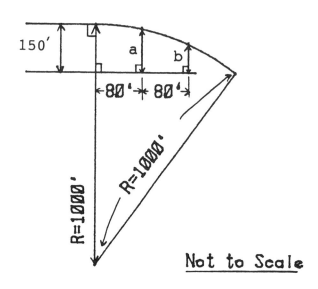

Not to Scale

Problem D
───────────

Line BCD in the Figure below represents a center line of
a survey for a certain part of a construction project.
Curve BC has a radius of 100.5 ft, with center at O.
Another construction line $\overline{AB}$, intersects line BCD at B.

REQUIRED
───────────

12. Find the coordinates of point O.

    (A) X = 159.92, Y = 20.82   (B) X = 0.0, Y = 0.0

    (C) X = 34.10, Y = 0.40   (D) X = 68.2, Y = 0.2

13. Find the length $\overline{OA}$ (ft).

    (F) 255.47   (G) 265.53   (H) 317.83   (J) 279.95

14. Find the angle $\angle$ BAO.

    (A) $14°12'35"$ (B) $14°35'12"$

    (C) $19°22'35"$ (D) $19°45'12"$

15. Find the angle $\angle$ ABO.

    (F) $112°30'40"$ (G) $52°19'45"$

    (H) $122°30'11"$ (J) $57°29'49"$

16. Find the Arc length BC (ft).

    (A) 126.58   (B) 129.12   (C) 121.38   (D) 122.74

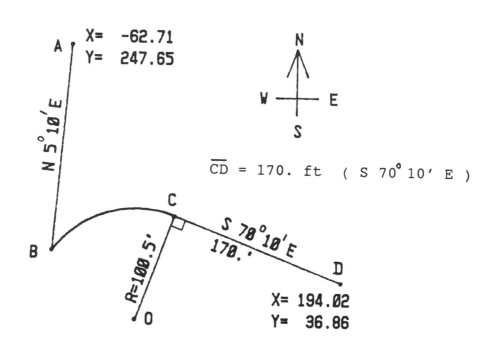

Problem E
_____

A reversed curve is composed of two simple curves turning
in opposite direction, as shown below. Both simple curves
have the same degree of curvature.

REQUIRED
_____

17. Find the radius (ft) of the simple curve.

   (A) 866.025   (B) 1082.53   (C) 2366.025   (D) 2957.532

18. Find the station of PC.

   (F) 18+66.03   (G) 16+33.98   (H) 20+00.00   (J) 16+66.03

19. Find the station of PRC.

   (A) 41+11.68   (B) 38+66.03   (C) 44+77.70   (D) 43+66.03

20. Find the station of PT.

   (F) 57+16.55   (G) 51+04.88   (H) 50+00.00   (J) 53+50.53

21. Find the station of PI2.

   (A) 36+33.98   (B) 47+16.55   (C) 47+45.65   (D) 50+00.00

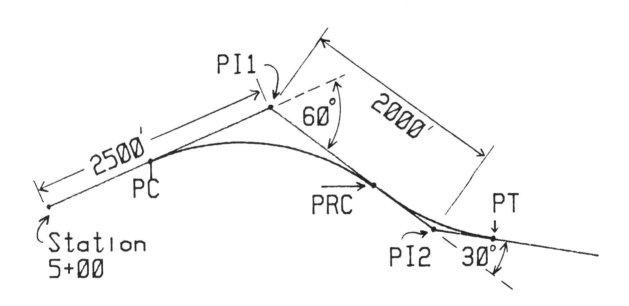

Problem F
_____

The tangent of a horizontal curve is relocated 20' away, and is parallel to the existing tangent as shown in the Figure below.

REQUIRED
_____

22. Find the radius of curvature R' for the new curve.

    (A) 966.35'   (B) 965.36'   (C) 956.36'   (D) 963.65'

23. What is the deflection angle from B.C. to the center of the new curve (Point A) ?

    (F) 30°   (G) 60°   (H) 15°   (J) 90°

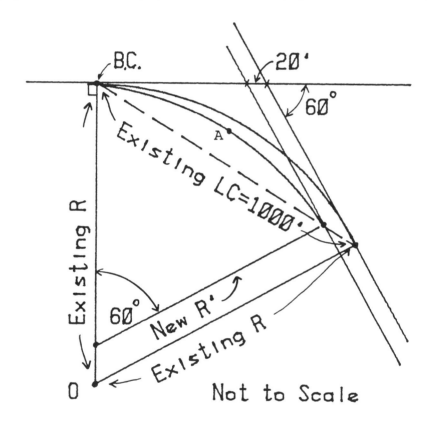

ANS :   1.A   2.G   3.C   4.H   5.D
        6.G   7.C   8.F   9.C   10.B
        11.F   12.B   13.F   14.C   15.H
        16.A   17.C   18.G   19.A   20.J
        21.D   22.B   23.H

1-23

2. VERTICAL CURVE

Vertical curves are used to provide a smooth transition between grade lines of a highway or railroad. An equal-tangent parabolic curve is illustrated in Fig. 2-A. It has the property that the vertex (V) is midway between the beginning of the vertical curve (BVC) and its end (EVC) measured horizontally. Parabolic curves can be calculated by two different procedures: Tangent-offset method or Chord-gradient method.

Tangent-offset method

Fig. 2-A shows: the center of a parabola is midway between the vertex and the long chord. Parabolic curve has the property that offsets from a tangent to a parabola are proportional to the squares of the horizontal distances from the point of tangency (See Eq.2-1).

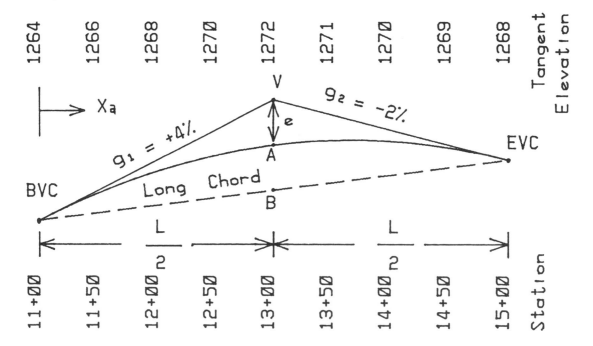

Figure 2-A

Terminology
_____

L	Horizontal Length of curve; in stations.
g_1 , g_2	Grade rates; in percentages.
BVC	Beginning of the Vertical Curve.
EVC	End of the Vertical Curve.
V	Vertex; point of intersection of two tangents.
r	Rate of change of grade; per station.
e	Middle ordinate
Xa	Horizontal distance from BVC (or EVC) to point a.

Table 2-1   Equal-Tangent Vertical Curve

Station	Point	Tangent Elevation	Tangent Offset	Elevation of Curve
11+00	BVC	1264	0.00	1264.00
11+50		1266	0.19	1265.81
12+00		1268	0.75	1267.25
12+50		1270	1.69	1268.31
13+00	V	1272	3.00	1269.00
13+50		1271	1.69	1269.31
14+00		1270	0.75	1269.25
14+50		1269	0.19	1268.81
15+00	EVC	1268	0.00	1268.00

1. Tangent offset

$$\frac{\text{Offset a}}{\text{Offset V}} = \left(\frac{Xa}{L/2}\right)^2 \qquad \text{(Eq.2-1)}$$

Calculate the tangent offset of Table 2-1.

$(1/4)^2$ x 3 = 0.19   (here, offset V = 3)

$(1/2)^2$ x 3 = 0.75

$(3/4)^2$ x 3 = 1.69

2. Middle ordinate $e = \dfrac{g_1 - g_2}{8} L$ (Eq.2-2)

From Fig.2-A, $e = \dfrac{4 - (-2)}{8} \ 4 = 3$

3. External distance $\overline{AB} = \dfrac{1}{2} \left(\dfrac{Y_{BVC} + Y_{EVC}}{2} - Y_v \right)$

From Fig.2-A, $\overline{AB} = \dfrac{1}{2} \left(\dfrac{1264 + 1268}{2} - 1272 \right) = 3$

4. Offset V = 3 = e = $\overline{AB}$ (ckecked)

$$\boxed{\text{Chord-gradient method}}$$

The following Equations define Equal-Tangent Vertical curve.

$$Y = Y_{BVC} + g_1 X + \frac{r X^2}{2} \qquad \text{(Eq.2-3)}$$

$$r = \frac{g_2 - g_1}{L} \qquad \text{(Eq.2-5)}$$

where Y is the curve elevation of any point P on the parabola, r is the rate of change of grade per station of the curve, g_1 the tangent grade through the BVC, and X is the horizontal distance from the BVC to point P. The HIGH or LOW point on a vertical curve will occur at the turning point where the tangent to the curve is horizontal and its slope is equal to zero. The turning point is not located directly above or below vertex (V). It is obtained by equating to zero the first derivative of Eq.2-3:

$$\frac{dY}{dX} = r X + g_1 = 0$$

$$\text{Turning point, } X = \frac{-g_1}{r} = \frac{g_1 (L)}{g_1 - g_2} \qquad \text{(Eq.2-4)}$$

For example, calculate X and Y of Figure 2-A:

$$r = \frac{g_2 - g_1}{L} = \frac{-2 - 4}{4} = -1.5$$

At station 13+00

$$X = (13+00) - (11+00) = 2+00$$

$$Y = Y_{BVC} + g_1 X + \frac{r X^2}{2} = 1264 + 4(2) + \frac{-1.5 (2)^2}{2} = 1269$$

At the turning point,

$$X = \frac{-g_1}{r} = \frac{g_1 (L)}{g_1 - g_2} = \frac{-4}{-1.5} = 2.667$$

$$Y = Y_{BVC} + g_1 X + \frac{r X^2}{2} = 1264 + 4(2.667) + \frac{-1.5 (2.667)^2}{2}$$

$$= 1269.33$$

The turning point is at station: $(11+00) + (2+66) = 13+66$

Example 2-1

A 600 ft vertical curve with grades g1 =(+)0.6% and g2 =(-)1.0%
that intersect at station 42+70 with elevation 820.00 ft. Find

(A) the stations and elevations of the BVC and EVC.

(B) the station and elevation at the high point of the curve.

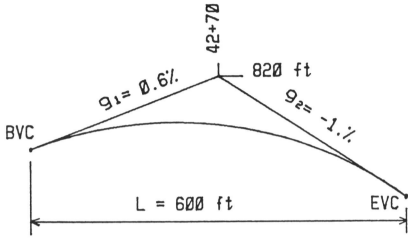

Solutions:

(A) BVC station:
   (42+70) - (L/2) = (42+70) - (3+00) = $\boxed{39+70}$

   BVC elevation:
   820 - g_1(L/2) = 820 - 0.6(3) = $\boxed{818.2\ ft}$

   EVC station:
   (42+70) + (L/2) = (42+70) + (3+00) = $\boxed{45+70}$

   EVC elevation:
   820 + g_2(L/2) = 820 - 1.0(3) = $\boxed{817.0\ ft}$

(B) The high point station:
   From (Eq.2-5), $r = \dfrac{g_2 - g_1}{L} = \dfrac{-1 -0.6}{6} = -0.267$

   $X = \dfrac{-g_1}{r} = \dfrac{g_1\ (L)}{g_1 - g_2} = \dfrac{-0.6}{-0.267} = 2.25$

   $Y = Y_{BVC} + g_1 X + \dfrac{r\ X^2}{2} = 818.2 + 0.6(2.25) + \dfrac{-0.267\ (2.25)^2}{2}$

   $= \boxed{818.87\ ft}$ (elevation)

   The high point is at station:

   (39+70) + (2+25) = $\boxed{41+95}$

2-4

Example 2-2
_____

A vertical curve with grades g1 =(-)0.8% and g2 =(+)1.0% that
start at station 42+70 with elevation 820.ft. At 48+70, there
is an overpass with an underside level of 835.2 ft and the
curve is designed for a 15-foot clearance under the overpass
at this point. Find the required curve length.

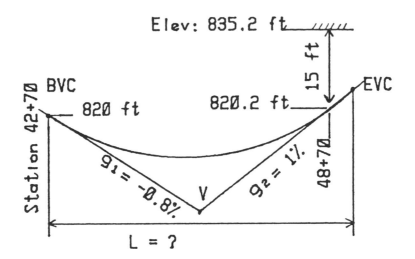

Solutions:
===============

(A) Method I (Chord-gradient method):

$$From\ (Eq.2\text{-}5),\quad r = \frac{g_2 - g_1}{L} = \frac{1 - (-0.8)}{L} = \frac{1.8}{L}$$

At station 48+70:

$$X = (48+70) - (42+70) = 6+00$$

$$Y = Y_{BVC} + g_1 X + \frac{r\ X^2}{2} = 820 + (-0.8)(6) + \frac{1}{2}(\frac{1.8}{L})\ (6)^2$$

$$= 835.2 - 15. = 820.2\ ft$$

$$\Rightarrow\quad 815.2 + (\frac{32.4}{L}) = 820.2$$

$$\Rightarrow\quad L = 6.48$$

The required curve length = $\boxed{648\ ft}$ *(longest)*

(B) Method II (Tangent-offset method):

    a. At station 48+70:

        $X = (48+70) - (42+70) = 6+00$

        $Y = 835.2 - 15. = 820.2$ *ft*

        Tangent elevation $= Y_{BVC} + g_1 X$

                        $= 820 - 0.8(6) = 815.2$ *ft*

        Tangent offset $= 820.2 - 815.2 = 5. ft$

    b. Tangent offset

$$\frac{Offset\ a}{Offset\ V} = (\frac{Xa}{L/2})^2 \qquad (Eq.2\text{-}1)$$

      Here, Offset V = e (Middle ordinate)

    c. Middle ordinate $\quad e = \dfrac{g_2 - g_1}{8} L = \dfrac{1-(-0.8)}{8} L = \dfrac{1.8}{8} L$

    d. Required curve length

$$\frac{Offset\ a}{Offset\ V} = (\frac{Xa}{L/2})^2 \qquad (Eq.2\text{-}1)$$

      Here, Offset V = e = 1.8 L/8

$$\Rightarrow \quad \frac{5}{1.8\ L/8} = (\frac{6}{L/2})^2$$

$$\Rightarrow \quad \frac{40}{1.8\ L} = \frac{144}{L^2}$$

$$\Rightarrow \quad L = 6.48$$

      The required curve length = $\boxed{648\ ft}$ *(longest)*

Example 2-3

A vertical curve with grades g1 =(-)3.0% and g2 =(+)2.0% that intersect at station 8+50 with elevation 1450.06 ft. It is necessary to pass under an overhead structure at station 8+80 with a 16-foot clearance. The lowest point of overhead structure is 1468.50 ft at station 8+80.

Determine (A) the longest length of the vertical curve.

   (B) the station and elevation of the low point on the curve.

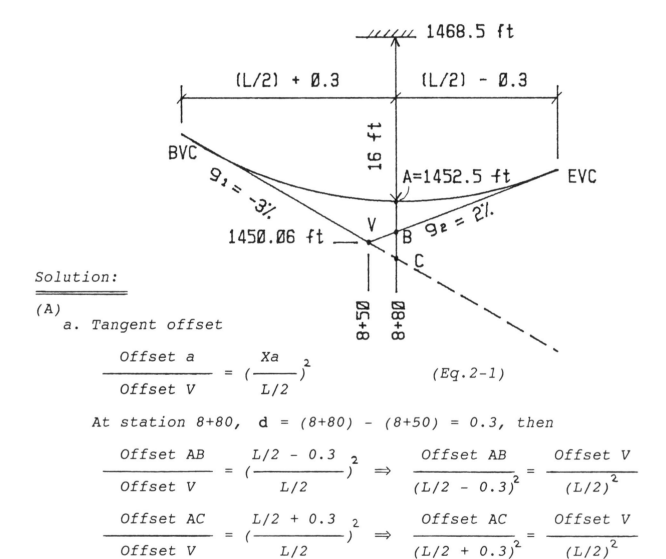

Solution:

(A)

   a. *Tangent offset*

$$\frac{Offset\ a}{Offset\ V} = (\frac{Xa}{L/2})^2 \qquad (Eq.2\text{-}1)$$

   At station 8+80, **d** = (8+80) - (8+50) = 0.3, then

$$\frac{Offset\ AB}{Offset\ V} = (\frac{L/2 - 0.3}{L/2})^2 \implies \frac{Offset\ AB}{(L/2 - 0.3)^2} = \frac{Offset\ V}{(L/2)^2}$$

$$\frac{Offset\ AC}{Offset\ V} = (\frac{L/2 + 0.3}{L/2})^2 \implies \frac{Offset\ AC}{(L/2 + 0.3)^2} = \frac{Offset\ V}{(L/2)^2}$$

Therefore, $\dfrac{\overline{AB}}{\left[(L/2) - 0.3\right]^2} = \dfrac{Offset\ V}{(L/2)^2} = \dfrac{\overline{AC}}{\left[(L/2) + 0.3\right]^2}$

(A)

b.

$$\frac{\overline{AB}}{\left[(L/2) - 0.3\right]^2} = \frac{\overline{AC}}{\left[(L/2) + 0.3\right]^2}$$

$\overline{AB} = 1452.5 - (1450.06 + 0.3 \times g_2) = 1.84'$

$\overline{AC} = 1452.5 - (1450.06 + 0.3 \times g_1) = 3.34'$

Substitue $\overline{AB} = 1.84'$, $\overline{AC} = 3.34'$ into the above equation,

$\Rightarrow 0.3 \ (\sqrt{1.84} + \sqrt{3.34}) = L \ (\sqrt{3.34} - \sqrt{1.84})/2$

$\Rightarrow L = 4.055$

The longest length of the vertical curve = $\boxed{405.5 \ ft}$

(B)

a. BVC station:
   $(8+50) - (L/2) = (8+50) - (2+2.75) = 6+47.25$

   BVC elevation:
   $1450.06 - g_1(L/2) = 1450.06 + 3(2.0275) = 1456.143 \ ft$

b. The low point station:
   From (Eq.2-5), $r = \dfrac{g_2 - g_1}{L} = \dfrac{2 - (-3)}{4.055} = 1.233$

   $X = \dfrac{-g_1}{r} = \dfrac{g_1 \ (L)}{g_1 - g_2} = \dfrac{-(-3)}{1.233} = 2.433$

   $Y = Y_{BVC} + g_1 X + \dfrac{r \ X^2}{2}$

   $= 1456.143 + (-3)(2.433) + \dfrac{1.233 \ (2.433)^2}{2}$

   $= \boxed{1452.49 \ ft}$ (elevation)

   The low point is at station:
   $(6+47.25) + (2+43.3) = \boxed{8+90.55}$

Example 2-4

A 300 ft vertical parabolic curve with grades g1 =(+)5.0%,
g2 =(-)2.5%, and the elevation of the high point on the curve
is 123.45 ft. Visibility is to be improved over this road
by replacing this curve by another 600 ft parabolic curve.
Find the depth of excavation required at the mid-point of
the curve.

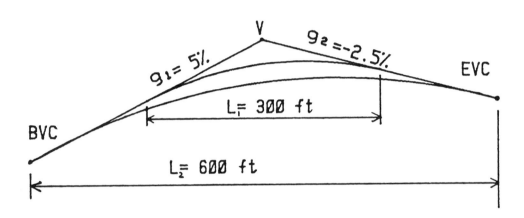

Solution:

$$\text{Offset } V = \text{Middle ordinate } e = \frac{g_1 - g_2}{8} L \qquad (Eq.2-2)$$

$$L_1 = 300.ft, \quad e_1 = \frac{g_1 - g_2}{8} L_1 = \frac{5 - (-2.5)}{8}(300) = 2.8125'$$

$$L_2 = 600.ft, \quad e_2 = \frac{g_1 - g_2}{8} L_2 = \frac{5 - (-2.5)}{8}(600) = 5.625'$$

*The depth of excavation required at the mid-point
of the curve:*

$$e_2 - e_1 = 5.625' - 2.8125' = \boxed{2.81 \text{ ft}}$$

Example 2-5

A 800 ft vertical curve with grades g1 =(+)2.0%, g2 =(-)3.0% and the data shown in the figure below.

(A) Find the station and elevation at the BVC, EVC, and the HIGH point on the curve.

(B) A new highway with an underside level 302 ft is running perpendicular to the existing curve. The new highway must maintain a 25-foot clearance over the existing curve below. Find the minimum and maximum station where the new highway should not be constructed.

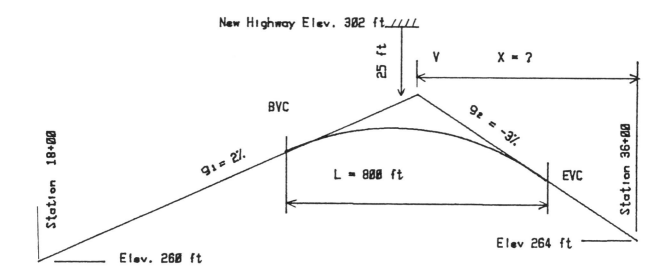

```
Station 18+00 : Elev.=260 ft
Station 36+00 : Elev.=264 ft
```

Solutions:

(A)
a. *Distance between Station (36+00) and (18+00) = 1800.ft*

Distance between V (Vertex) and Station 36+00 = X

Elevation from station (18+00) to (36+00):

$$Y_{18+00} + g_1 (1800 - X) + g_2 (X) = Y_{36+00}$$

$$\Rightarrow 260 + 0.02(1800 - X) - 0.03(X) = 264$$

$$\Rightarrow X = 640.ft$$

Vertex station:
(36+00) - X = (36+00) - (6+40) = 29+60

Vertex elevation:
$264 - g_2 X = 264 + 0.03(640) = 283.2$ ft

BVC station:
(29+60) - (L/2) = (29+60) - (4+00) = $\boxed{25+60}$

BVC elevation:
$283.2 - g_1 (L/2) = 283.2 - 2(4) = \boxed{275.2 \text{ ft}}$

EVC station:
(29+60) + (L/2) = (29+60) + (4+00) = $\boxed{33+60}$

EVC elevation:
$283.2 + g_2 (L/2) = 283.2 - 3(4) = \boxed{271.2 \text{ ft}}$

b. *The high point station:*

$$\text{From (Eq.2-5), } \quad r = \frac{g_2 - g_1}{L} = \frac{-3 - 2}{8} = -0.625$$

$$X = \frac{-g_1}{r} = \frac{g_1 (L)}{g_1 - g_2} = \frac{-2}{-0.625} = 3.2$$

$$Y = Y_{BVC} + g_1 X + \frac{r X^2}{2} = 275.2 + 2(3.2) + \frac{-0.625 (3.2)^2}{2}$$

$$= \boxed{278.4 \text{ ft}} \quad \text{(elevation)}$$

The high point is at station:

$$(25+60) + (3+20) = \boxed{28+80}$$

2-11

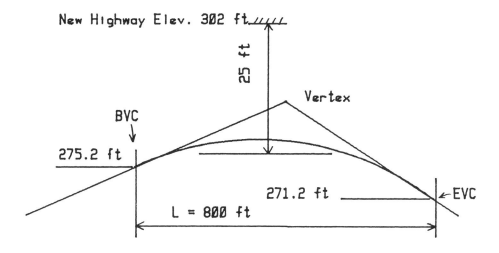

New Highway Elev. 302 ft

25 ft

Vertex

BVC

275.2 ft

271.2 ft

EVC

L = 800 ft

(B) New highway location:

Maintain a 25-foot clearance:

$Y = 302 - 25 = 277$ ft

$$Y = Y_{BVC} + g_1 X + \frac{r\ X^2}{2} = 275.2 + 2(X) + \frac{-0.625\ (X)^2}{2}$$

$= 277.$ ft

$\Rightarrow$ $X = 1.083'$ or $5.317'$

The new highway should not be constructed between:

Station: $(25+60) + (1+8.3) = \boxed{26+68.3}$ (Min)

$(25+60) + (5+31.7) = \boxed{30+91.7}$ (Max)

Sample Problems 2

Problem A

_____

A vertical curve with grades g1 =(+)0.8% and g2 =(-)0.4%
that intersect at station 18+00 with elevation 820.00 ft.
The maximum allowable change in grade per station is 0.2.

REQUIRED

_____

1. What is the minimum length of a vertical connecting
   this two grades ?

   (A) 200'   (B) 400'   (C) 600'   (D) 800'

2. Find the elevation (ft) of the BVC.

   (F) 817.6   (G) 818.8   (H) 821.2   (J) 822.4

3. Find the elevation of the mid-point of the long chord.

   (A) 817.8'   (B) 818.2'   (C) 821.2'   (D) 819.1'

4. Find the elevation (ft) of the mid-point of the vertical
   curve between the BVC and EVC.

   (F) 817.8   (G) 818.2   (H) 821.2   (J) 819.1

5. Find the station of the summit of the curve.

   (A) 17+50   (B) 18+50   (C) 19+00   (D) 19+50

6. Find the elevation (ft) of the summit of the curve.

   (F) 819.0   (G) 818.2   (H) 821.2   (J) 819.2

7. Find the middle ordinate (ft) of the curve.

   (A) 0.3   (B) 0.6   (C) 0.9   (D) 1.2

8. Find the tangent offset (ft) at station 17+00.

   (F) 0.1   (G) 0.4   (H) 0.6   (J) 0.9

Problem B

A vertical curve with grades g1 =(+)1.25% and g2 =(-)2.75%
that intersect at station 8+00 with elevation 820.00 ft.
The length of the curve is to be 600 ft.

REQUIRED
---

9. Find the elevation (ft) of the BVC.

   (A) 823.75   (B) 818.125   (C) 816.25   (D) 811.75

10. Determine the rate of change of grade per station (%).

   (F) +0.667   (G) -0.25   (H) +0.25   (J) -0.667

11. Find the station at the high point of the curve.

   (A) 6+87.5   (B) 6+78.5   (C) 8+00   (D) 9+12.5

12. Find the elevation (ft) at the high point of the curve.

   (F) 817.25   (G) 817.42   (H) 818.25   (J) 818.42

ANS :   1.C   2.F   3.B   4.J   5.C
         6.J   7.C   8.G   9.C   10.J
        11.A   12.G

3. TRAVERSE

A traverse is a series of consecutive lines or courses on which the lengths and directions have been determined. A traverse which comes back to its starting point is called a closed traverse. Closed traverses provide checks on the angles and distances. An open traverse is a series of lines which do not return to the starting point. Open traverses are sometimes used on route survey.

Meridian: The direction of a line may be expressed as an angle from an established reference line. This reference line is called reference meridian and may be expressed as one of the following:

    1. Magnetic meridian (influenced by magnetic pole).

    2. True meridian (passing the north and south pole).

    3. Assumed meridian (arbitrarily chosen).

Bearing: The bearing of a line is the ACUTE horizontal angle between a reference meridian and the line.

Bearing $\overline{AB}$: N 30° E

Bearing $\overline{BA}$: S 30° W

Back Bearing of $\overline{AB}$: S 30° W

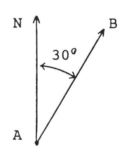

Azimuth: Azimuth is angle measured clockwise from a reference meridian.

Azimuth is from 0° to 360°.

Azimuth $\overline{AB}$ (from the south): 30°

Azimuth $\overline{AB}$ (from the north): 210°

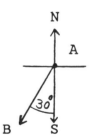

Back Azimuth $\overline{AB}$ (from the north): 30°

Latitude: The projection of a course on the north-south direction.

Departure: The projection of a course on the east-west direction.

$$\text{Latitude} = L \, (\cos \alpha)$$

$$\text{Departure} = L \, (\sin \alpha)$$

L = length $\overline{AB}$

α = bearing or azimuth

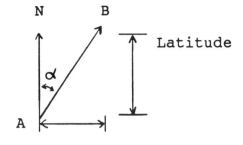

Angle Adjustment

The first step in traverse calculation is to balance the angles. The sum of the interior angles of a polygon with n sides is (n - 2) X (180°). Angles of a closed traverse can be adjusted to the correct geometric total by arbitrary correction or an average correction (even distribution). Unlike correction for linear measurement, the angle adjustment is independent of the size of an angle. An example is illustrated below.

Point	Measured Angle	Arbitrary Correction		Average Correction	
		Adjust-ment	Adjusted Angle	Adjust-ment	Adjusted Angle
A	105°20'	5'	105°15'	1'	105°19'
B	86°56'	0	86°56'	1'	86°55'
C	82°48'	0	82°48'	1'	82°47'
D	159°31'	0	159°31'	1'	159°30'
E	105°30'	0	105°30'	1'	105°29'
Total	540°05'	5'	540°00'	5'	540°00'

Table 3-1

Traverse Adjustment

There are five methods for traverse adjustment:

        (1) Compass rule.
        (2) Least squares method.
        (3) Arbitrary method.
        (4) Transit rule.
        (5) Crandall method.

The compass rule (or Bowditch rule) is the most commonly used method in practice. It is suitable for survey where the angles and the distances are measured with equal precision. The other four methods are seldom used.

Using the compass rule, corrections are made as follows:

$$\frac{\text{correction in latitude of leg}}{\text{closure in latitude}} = \frac{\text{length of leg}}{\text{traverse perimeter}}$$

$$\frac{\text{correction in departure of leg}}{\text{closure in departure}} = \frac{\text{length of leg}}{\text{traverse perimeter}}$$

The linear error of closure of a traverse is as follows:

Linear error of closure

$$= \sqrt{(\text{closure in departure})^2 + (\text{closure in latitude})^2}$$

The relative error of closure (or precision) for a traverse is expressed by

$$\text{Precision} = \frac{\text{linear error of closure}}{\text{traverse perimeter length}}$$

An example for traverse adjustment is illustrated below:

SITUATION:

The data obtained in a partially completed survey of an industrial building site are giving in Figure 3-A below.

REQUIREMENTS:

(A) Compute the bearing of lines BC, CD, DE, and EA.

(B) Calculate the corrections and precision for traverse adjustment.

Solutions:

(A) Compute the bearing of lines BC, CD, DE, and EA

Line	Bearing
BC	S 51°26'E
CD	S 41°38'W
DE	N 41°10'W
EA	N 20°41'W

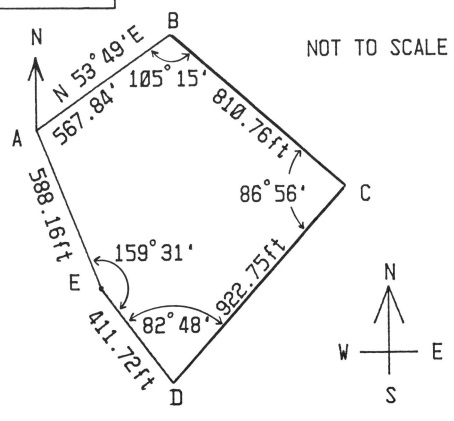

Figure 3-A

(B) Traverse Adjustment (By Compass rule)

Line	Bearing	Distance	Latitude (cos)	Departure (sin)
AB	N 53°49'E	567.84	335.24	458.32
BC	S 51°26'E	810.76	-505.45	633.92
CD	S 41°38'W	922.75	-689.67	-613.04
DE	N 41°10'W	411.72	309.94	-271.02
EA	N 20°41'W	588.16	550.25	-207.74
Total		3301.23	+0.31	+0.44

$$M = -\frac{0.31}{3301.23} \qquad N = -\frac{0.44}{3301.23}$$

Line	Corrections		Adjusted	
	Latitude	Departure	Latitude	Departure
AB	M × AB = -0.05	N × AB = -0.08	335.19	458.24
BC	M × BC = -0.08	N × BC = -0.11	-505.53	633.81
CD	M × CD = -0.09	N × CD = -0.12	-689.76	-613.16
DE	M × DE = -0.04	N × DE = -0.05	309.90	-271.07
EA	M × EA = -0.05	N × EA = -0.08	550.20	-207.82
Total	-0.31	-0.44	0.0	0.0

$$\text{Linear error of closure} = \sqrt{0.31^2 + 0.44^2} = 0.54 \text{ ft}$$

$$\text{Precision} = \frac{0.54}{3301.23} = \frac{1}{6113} \cong \frac{1}{6200}$$

3-5

Example 3-1

What is the bearing of each side of Figure 3-B?

Given: 1 = 110°15'15"
        2 =   80°53'02"
        3 =   89°01'45"
        4 =   79°49'58"

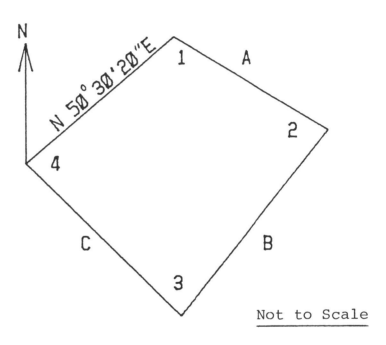

Figure 3-B

Solution:

Check the sum of the interior angles:

$110^{\circ}15'15'' + 80^{\circ}53'02'' + 89^{\circ}01'45'' + 79^{\circ}49'58'' = 360^{\circ}0'0''$

$180^{\circ} \times (n - 2) = 180^{\circ} \times (4 - 2) = 360^{\circ}$ (Checked)

Side	Line	Bearing
A	12	S $59^{\circ}44'55''$E
B	23	S $39^{\circ}22'03''$W
C	34	N $49^{\circ}39'42''$W

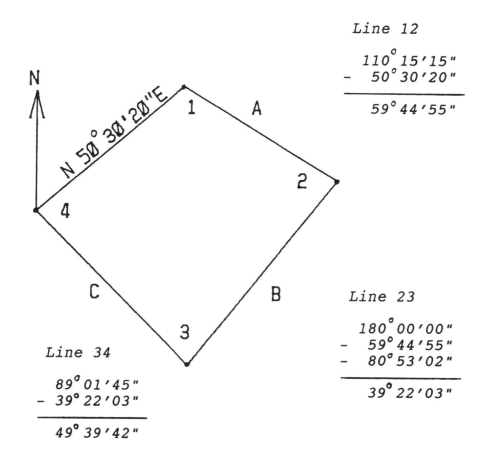

Line 12

$\begin{array}{r} 110^{\circ}15'15'' \\ -\ 50^{\circ}30'20'' \\ \hline 59^{\circ}44'55'' \end{array}$

Line 23

$\begin{array}{r} 180^{\circ}00'00'' \\ -\ 59^{\circ}44'55'' \\ -\ 80^{\circ}53'02'' \\ \hline 39^{\circ}22'03'' \end{array}$

Line 34

$\begin{array}{r} 89^{\circ}01'45'' \\ -\ 39^{\circ}22'03'' \\ \hline 49^{\circ}39'42'' \end{array}$

3-7

Example 3-2

A contractor has driven a tunnel heading from point B to point A, and another from point B to point D. He has requested you, as the resident engineer of the construction contract, to provide him with the correct bearing and distance so that he can connect points A and D with a connecting tunnel.

Points A and D are connected by a surface traverse as follows:

Course	Bearing	Distance (level)	Elevations
AB	S 12°14'E	1291.30 ft	At A, 357.6 ft
BC	N 83°12'E	1317.40 ft	At B, 454.7 ft
CD	S 89°41'E	1819.80 ft	At D, 892.4 ft

REQUIRED:

(A) What is the bearing of the connecting tunnel A-D?

(B) What is the length of the connecting tunnel A-D?

(C) What is the slope of the connecting tunnel A-D?

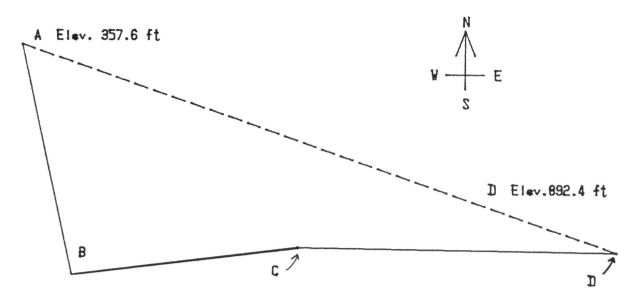

Solutions:
=========

Using point A as the origin (0,0),
calculate the coordinates as follows:

L.	Bearing	Dista.	Latitu. (cos)	Depart. (sin)	P.	Y	X
					A	0.0	0.0
AB	S 12°14'E	1291.3	-1261.98	273.62	B	-1261.98	273.62
BC	N 83°12'E	1317.4	155.99	1380.13	C	-1105.99	1581.75
CD	S 89°41'E	1819.8	-10.06	1819.77	D	-1116.05	3401.52

(A) Bearing $\overline{AD}$ = $\tan^{-1}\dfrac{3401.52}{-1116.05}$ = $\boxed{\text{S } 71°50'6.6''E}$

(B) Distance $\overline{AD}$ = $\sqrt{1116.05^2 + 3401.52^2}$ = 3579.93 ft (level)

Vertical distance = 892.4 - 357.6 = 534.8 ft

Length $\overline{AD}$ = $\sqrt{3579.93^2 + 534.8^2}$ = $\boxed{3619.66 \text{ ft}}$

(C) Slope = $\dfrac{534.8}{3579.93}$ = $\boxed{14.94\%}$

Example 3-3
_____

You are provided, by an individual authorized to practice land
surveying, the following property description:

Beginning at most westerly corner of Lot 1, Tract No.2 as shown
below.
        Thence N 55°37'E   567.34 feet;
        thence S 49°38'E   810.76 feet;
        thence S 43°26'W   922.75 feet;
        thence N 39°22'W   411.72 feet;
        thence N 18°53'W   588.16 feet
        to the point of beginning.

REQUIRED
_____

(A) Calculate all interior angles.

(B) Check the sum of the interior angles.

(C) Calculate the deflection angle on each station.

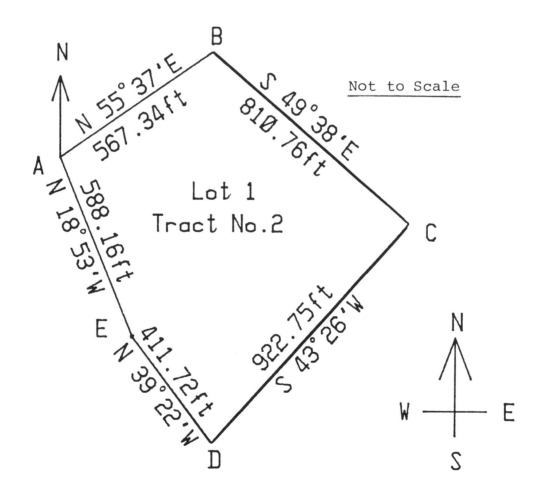

Solutions:

(A) The interior angles:

∠ A

    180° 00′
  − 55° 37′
  − 18° 53′
    ―――――――
    105° 30′

∠ B

    55° 37′
  + 49° 38′
   ――――――――
    105° 15′

∠ C

    180° 00′
  − 49° 38′
  − 43° 26′
   ――――――――
    86° 56′

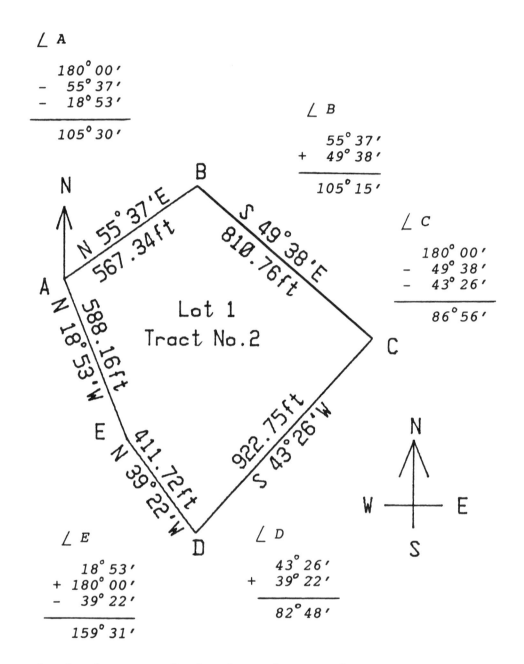

∠ E

    18° 53′
  + 180° 00′
  − 39° 22′
   ――――――――
    159° 31′

∠ D

    43° 26′
  + 39° 22′
   ――――――――
    82° 48′

(B) Check the sum of the interior angles:

105° 15′ + 105° 30′ + 159° 31′ + 82° 48′ + 86° 56′ = 540° 0′

180° x (n − 2) = 180° x (5 − 2) = 540° (Checked)

(C) Show the deflection angles.

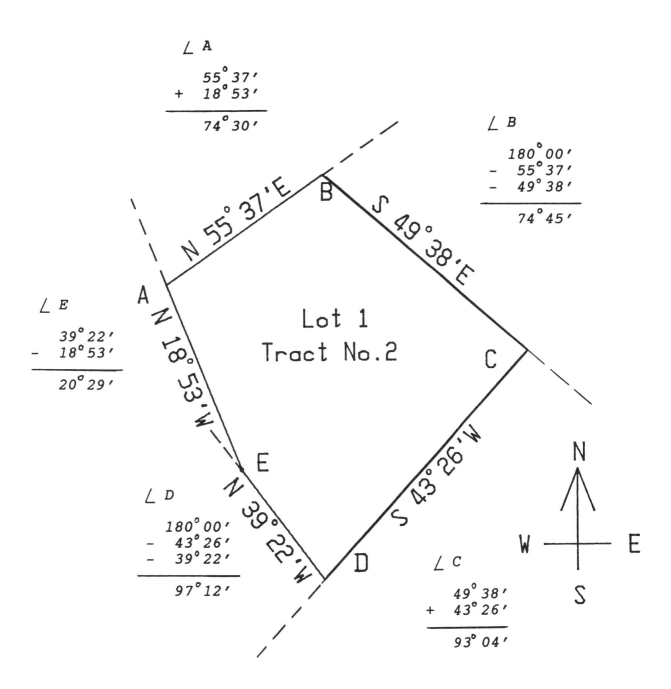

∠ A

$$55°37'$$
$$+ \; 18°53'$$
$$74°30'$$

∠ B

$$180°00'$$
$$- \; 55°37'$$
$$- \; 49°38'$$
$$74°45'$$

∠ E

$$39°22'$$
$$- \; 18°53'$$
$$20°29'$$

∠ D

$$180°00'$$
$$- \; 43°26'$$
$$- \; 39°22'$$
$$97°12'$$

∠ C

$$49°38'$$
$$+ \; 43°26'$$
$$93°04'$$

N 55° 37' E

S 49° 38' E

N 18° 53' W

S 43° 26' W

N 39° 22' W

Lot 1
Tract No.2

A   B   C   D   E

N
W   E
S

Example 3-4

The following figures show the double verniers. Determine the reading for the two sets of numbers on each circle.

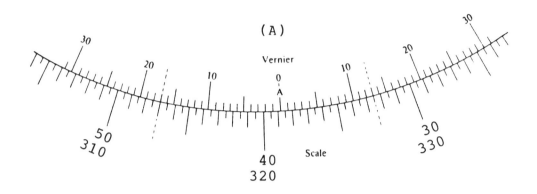

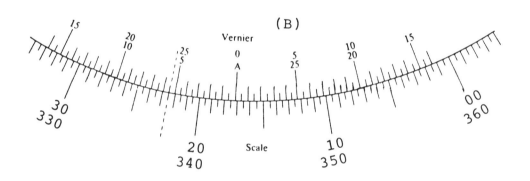

Solution:

Figure (A):
   The reading for the inner set is $38°30' + 17' = 38°47'$

   The reading for the outer set is $321°00' + 13' = 321°13'$

Figure (B):
   The reading for the inner set is $17° + 05'30" = 17°05'30"$

   The reading for the outer set is $343° - 05'30" = 342°54'30"$

Sample Problems 3

Problem A

Two separate traverses start from Station A, which is located at the portal of a tunnel. Traverse #1 was run in the tunnel, and traverse #2 was run on the surface. The field measurements are reproduced as follows:

Note: Horizontal distances and north azimuths are used throughout.

Tunnel Traverse # 1:

Station A to Station B, Azimuth = 300°00',
Distance = 300.00 feet, Grade = +1 per cent

Station B to breast of tunnel, Azimuth = 30°00',
Distance = 100.00 feet, Grade = +2 per cent

Tunnel Traverse # 2:

Station A to Station C, Azimuth = 30°00',
Distance = 200.00 feet, Grade = +2 per cent

Station C to Station D, Azimuth = 330°00',
Distance = 300.00 feet, Grade = +25 per cent

A vertical shaft is to be sunk at Station D, and the breast of the tunnel is to be connected with the shaft by a drift having a +3 per cent grade. (+3% from breast of the tunnel)

REQUIRED

1. Determine the azimuth of the drift

   (A) 39°38'   (B) 39°08'   (C) 38°08'   (D) 39°48'

2. Determine the required depth of shaft (ft).

   (F) 66.40   (G) 64.60   (H) 68.60   (J) 60.40

3. Determine the slope length of the drift (ft).

   (A) 250.33   (B) 258.33   (C) 255.53   (D) 253.33

---

4. If the coordinates of Point A are 1025 N, 1575 W and those of point B are 425 N, 975 W, the bearing of course AB is :

   (F) N 45°E   (G) S 45°W   (H) S 45°E   (J) N 45°W

Problem B

You are provided, by an individual authorized to practice land surveying, the following property description:

Beginning at most westerly corner of Lot 1, Tract No.2 as shown below.

Thence N 37°04'E	514.1	feet;
thence S 68°32'E	1395.6	feet;
thence S 55°40'W	961.3	feet;
thence N 32°48'W	243.3	feet;
thence N 57°17'W	816.5	feet

to the point of beginning.

REQUIRED
---

5. Calculate the interior angle at point A.

   (A) 37°04'   (B) 85°39'   (C) 57°17'   (D) 94°21'

6. Calculate the sum of the interior angles.

   (F) 539°00'   (G) 539°30'   (H) 540°00'   (J) 540°30'

7. Find the linear error of closure of the traverse by the Compass rule.

   (A) 5.0'   (B) 5.5'   (C) 4.5'   (D) 6.0'

8. Find the precision (relative error of closure) of the traverse by the Compass rule.

   (F) 1/786   (G) 1/714   (H) 1/873   (J) 1/655

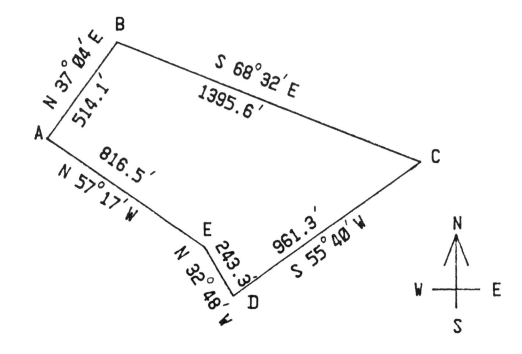

9. A backsight with inverted telescope is sighted at a point that bears N 38° E from the transit. The telescope is plunged (reversed) and a right deflection angle of 68° is turned. The bearing of the new line is :

   (A) N 74° W   (B) N 16° W   (C) S 16° E   (D) S 74° E

10. When set up at the B.C. (Beginning of Curve) of a simple curve, the angle to turn off the P.I. (Point of Intersection) to hit the radius point is :

    (F) one half delta   (G) delta

    (H) one radian   (J) 90°

11. The following figures show the transit verniers.

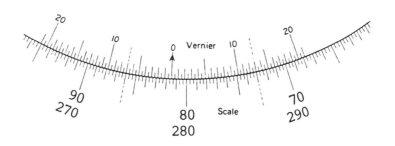

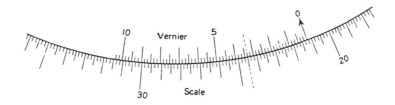

   For a clockwise angle, determine the reading of the double vernier.

   (A) 86° 07'   (B) 81° 07'   (C) 81° 27'   (D) 73° 13'

   For a clockwise angle, determine the reading of the direct vernier.

   (F) 24° 13' 20"   (G) 21° 13' 20"

   (H) 24° 30' 00"   (J) 21° 03' 20"

ANS : 1.B   2.F   3.D   4.H   5.B   6.H
      7.A   8.F   9.A   10.J   11.C G

4. AREA

The area of a closed traverse can be calculated by the Double-Meridian-Distance (DMD) method. Before calculating the area, the traverse should be checked and adjusted for closure.

The meridian distance of a traverse course is the perpendicular distance from the center point of the course to the reference meridian. To ease the signs problem, the reference meridian usually is placed through the most westerly traverse station. The following rules can be applied in calculating DMD:

1. DMD of the first course = departure of the first course.

2. DMD for any course = DMD of the preceding course

    + departure of the preceding course

    + departure of the course itself.

3. DMD of the last course = (-) departure of the last course.

The area is calculated by the following equation:

$$A = \frac{1}{2} \left| \sum_{i=1}^{n} (latitude_i) \times (DMD_i) \right|$$

and 1 acre = 43,560 ft^2, 1 mile = 5,280 ft.

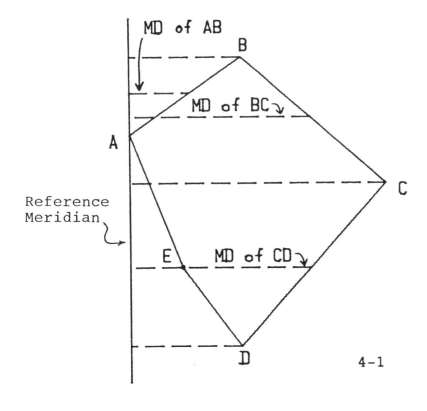

An example for the DMD method is illustrated as follows:

You are provided, by an individual authorized to practice land surveying, the following property description:

Beginning at most westerly corner of Lot 1, Tract No.2 as shown below.

Thence N 55°37′E	567.34	feet;
thence S 49°38′E	810.76	feet;
thence S 43°26′W	922.75	feet;
thence N 39°22′W	411.72	feet;
thence N 18°53′W	588.16	feet

to the point of beginning.

REQUIRED

Calculate the area ABCDEA.

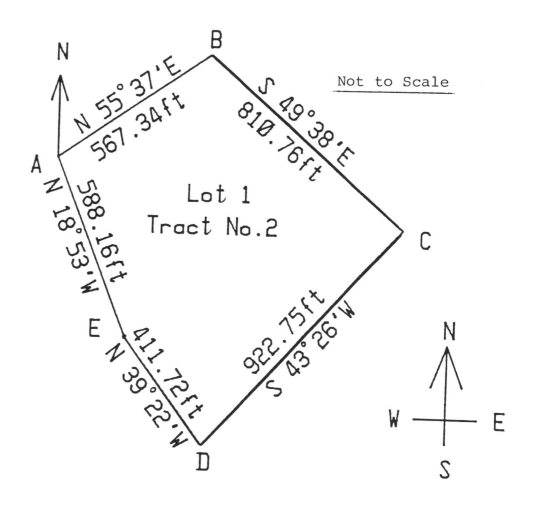

Solution:

(A) Check the sum of the interior angles as in example 3-3.

(B) Calculate the area ABCDEA by the DMD method

Line	Bearing	Distance	Latitude (cos)	Departure (sin)	Correction L.	Correction D.
AB	N 55°37′E	567.34	320.39	468.21	0.	-0.01
BC	S 49°38′E	810.76	-525.11	617.73	0.	-0.01
CD	S 43°26′W	922.75	-670.08	-634.40	0.	-0.01
DE	N 39°22′W	411.72	318.30	-261.15	0.	-0.00
EA	N 18°53′W	588.16	556.50	-190.35	0.	-0.01
			0.0	+0.04	0.	-0.04

Line	(Adjusted) Latitude	Departure	DMD	Lat. × DMD
AB	320.39	468.20	468.20	150006.60
BC	-525.11	617.72	1554.12	-816083.95
CD	-670.08	-634.41	1537.43	-1030201.09
DE	318.30	-261.15	641.87	204307.22
EA	556.50	-190.36	190.36	105935.34
				-1386035.89

$$\text{Area} = \frac{1}{2} \left(\frac{1386035.89}{43560} \right) = 15.91 \text{ acres}$$

$$\boxed{\text{Coordinates} \quad \text{Method}}$$

The area of a closed traverse can be calculated by the coordinates method with known coordinate for each corner.

$$2 \times (A) = \left[\sum_{i=1}^{n} Y_i (X_{i-1} - X_{i+1}) \right]$$

The above equation can be easily remembered by listing the X and Y coordinates as follows: (for area ABC)

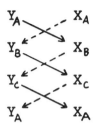

The starting point repeated at the end, the solid arrows considered (−), the dashed arrows considered (+). The above example illustrated for DMD method is re-calculated here by the coordinates method.

Line	Latitude	Departure	Station	Y	X
			A	320.39	468.20
AB	320.39	468.20			
			B	-204.72	1085.92
BC	-525.11	617.72			
			C	-874.80	451.51
CD	-670.08	-634.41			
			D	-556.50	190.36
DE	318.30	-261.15			
			E	0.0	0.0
EA	556.50	-190.36			
			A	320.39	468.20

$$\text{Area} = \frac{1}{2} \left(\frac{1}{43560}\right) \times$$

$$\left[\begin{array}{l} - (320.39)(1085.92) - (-204.72)(451.51) \\ - (-874.80)(190.36) + (468.20)(-204.72) \\ + (1085.92)(-874.80) + (451.51)(-556.50) \end{array} \right]$$

$$= \frac{1}{2} \left(\frac{1386035.89}{43560}\right) = \boxed{15.91 \text{ acres}}$$

For the area of a triangle, Caltrans Surveys are re-printed on the next two pages.

3. RIGHT TRIANGLES

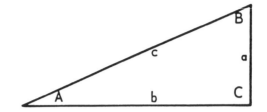

$C = 90°$

GIVEN	TO FIND	FORMULAS
a, c	b A B	$\sqrt{c^2 - a^2}$ $\sin A = a/c$ $\cos B = a/c$
	Area	$\frac{a}{2}\sqrt{c^2 - a^2}$
b, c	a A B	$\sqrt{c^2 - b^2}$ $\cos A = b/c$ $\sin B = b/c$
	Area	$\frac{b}{2}\sqrt{c^2 - b^2}$
a, b	c A B	$\sqrt{a^2 + b^2}$ $\tan A = a/b$; $\cot A = b/a$ $\tan B = b/a$; $\cot B = a/b$
	Area	$ab/2$
A, a	b c B	$a \cot A$ $a/\sin A$ $90° - A$
	Area	$\dfrac{a^2 \cot A}{2}$
A, b	a c B	$b \tan A$ $b/\cos A$ $90° - A$
	Area	$\dfrac{b^2 \tan A}{2}$
A, c	a b B	$c \sin A$ $c \cos A$ $90° - A$
	Area	$\dfrac{c^2(\sin A)(\cos A)}{2} = \dfrac{c^2 \sin 2A}{4}$

D. BASIC TRIGONOMETRY

4. OBLIQUE TRIANGLES

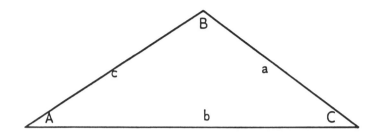

Law of Sines: $\dfrac{a}{\sin A} = \dfrac{b}{\sin B} = \dfrac{c}{\sin C}$

Law of Cosines: $a^2 = b^2 + c^2 - 2bc \cos A$

or $\cos A = \dfrac{b^2 + c^2 - a^2}{2bc}$

GIVEN	TO FIND	FORMULAS
a, b, c	A, B, & C using "s"	Law of Cosines, $\sin \tfrac{1}{2} A = \sqrt{\dfrac{(s-b)(s-c)}{bc}}$; $\cos \tfrac{1}{2} A = \sqrt{\dfrac{s(s-a)}{bc}}$ $\sin A = \dfrac{2\sqrt{s(s-a)(s-b)(s-c)}}{bc}$ *NOTE: For angles B&C make appropriate substitutions in these formulas* The value "s" = $\tfrac{1}{2}$(a+b+c)
	Area	$\sqrt{s(s-a)(s-b)(s-c)}$ The value "s" = $\tfrac{1}{2}$(a+b+c)
a, A, B	b C c	Law of Sines $180° - (A + B)$ Law of Sines ; $\dfrac{a \sin(A+B)}{\sin A}$
	Area	$\dfrac{a^2 \, \sin B \, \sin(A+B)}{2 \sin A}$
a, b, A	B C c	Law of Sines $180° - (A + B)$ $\dfrac{a \sin (A + B)}{\sin A}$; Law of Sines
a, b, C	c A B	Law of Cosines $\tan A = \dfrac{a \sin C}{b-(a\cos C)}$ $180° - (A + C)$
	Area	$\tfrac{1}{2}$ ab $\sin C$
ABC, a	Area	$\dfrac{a^2 (\sin B)(\sin C)}{2 \sin A}$

Example 4-1

The data obtained in a partially completed survey of an industrial building site are giving in Figure 4-A below.

REQUIREMENTS:

(A) Compute the bearing of lines BC, CD, DE, and EA.

(B) Compute the interior angle at point A.

(C) Compute the area ABCDEA (in acres).

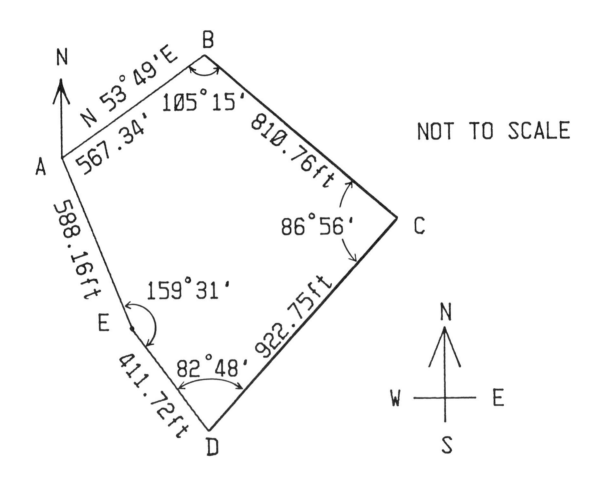

Figure 4-A

Solutions:

(A) *Compute the bearing of lines BC, CD, DE, and EA*

Line	Bearing
BC	S 51°26'E
CD	S 41°38'W
DE	N 41°10'W
EA	N 20°41'W

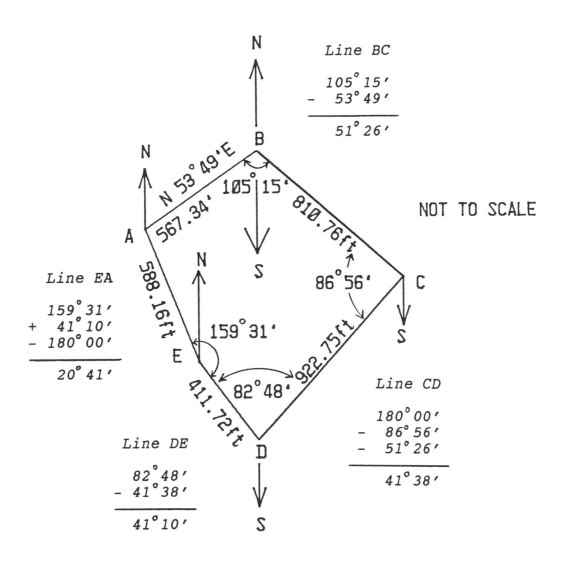

Line BC

105° 15'
−   53° 49'
─────────
51° 26'

NOT TO SCALE

Line EA

159° 31'
+   41° 10'
− 180° 00'
─────────
20° 41'

Line CD

180° 00'
−   86° 56'
−   51° 26'
─────────
41° 38'

Line DE

82° 48'
− 41° 38'
─────────
41° 10'

(B) The sum of the interior angles should be

$$(n - 2) \times 180^0 = (5 - 2) \times 180^0 = 540^0$$

The interior angle at point A:

$$540^0 - (105^0 15' + 86^0 56' + 82^0 48' + 159^0 31') = 105^0 30'$$

(C) Method I

Calculate the area ABCDEA by the DMD method
(Double Meridian Distance)

Line	Bearing	Distance	Latitude (cos)	Departure (sin)	Correction L.	Correction D.
AB	N 53°49'E	567.34	334.94	457.92	0.	-0.01
BC	S 51°26'E	810.76	-505.45	633.92	0.	-0.01
CD	S 41°38'W	922.75	-689.67	-613.04	-0.01	-0.01
DE	N 41°10'W	411.72	309.94	-271.02	0.	0.00
EA	N 20°41'W	588.16	550.25	-207.74	0.	-0.01
			+0.01	+0.04	-0.01	-0.04

Line	(Adjusted) Latitude	Departure	DMD	Lat. × DMD
AB	334.94	457.91	457.91	153372.38
BC	-505.45	633.91	1549.73	-783311.03
CD	-689.68	-613.05	1570.59	-1083204.51
DE	309.94	-271.02	686.52	212780.01
EA	550.25	-207.75	207.75	114314.44
				-1386048.71

$$\text{Area} = \frac{1}{2} \left(\frac{1386048.71}{43560} \right) = 15.91 \text{ acres}$$

(C) Method II

Calculate the area ABCDEA by the Coordinates Method.

Line	Latitude	Departure	Station	Y	X
AB	334.94	-457.91	A	334.94	457.91
			B	-170.51	1091.82
BC	-505.45	633.91			
			C	-860.19	478.77
CD	-689.68	-613.05			
			D	-550.25	207.75
DE	309.94	-271.02			
			E	0.0	0.0
EA	550.25	-207.75			
			A	334.94	457.91

$$
Area = \frac{1}{2} \left(\frac{1}{43560} \right) \times
$$

$$
\left[
\begin{array}{l}
- (334.94)(1091.82) - (-170.51)(478.77) \\
- (-860.19)(207.75) + (457.91)(-170.51) \\
+ (1091.82)(-860.19) + (478.77)(-550.25)
\end{array}
\right]
$$

$$
= \frac{1}{2} \left(\frac{1386048.72}{43560} \right)
$$

$$
= \boxed{15.91 \ acres}
$$

Example 4-2

Determine the area BEC in Figure 4-B

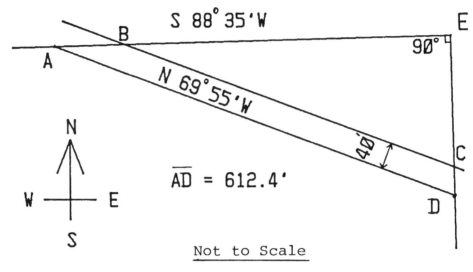

Not to Scale

Figure 4-B

Solution:

a. $\angle EBC = 180^0 - 69^055' - 88^035' = 21^030'$

$\overline{CE} = \overline{DE} - \overline{DC}$

$\qquad = 612.4 (sin\ 21.5^0) - \dfrac{40}{cos\ 21.5^0}$

$\qquad = 181.45'$

$\overline{BE} = \overline{AE} - \overline{AB}$

$\qquad = 612.4 (cos\ 21.5^0) - \dfrac{40}{sin\ 21.5^0}$

$\qquad = 460.65'$

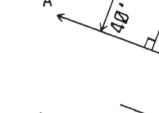

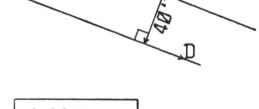

b. Area BEC $= \dfrac{1}{2}(\overline{CE} \times \overline{BE})$

$\qquad = \dfrac{1}{2}(\dfrac{181.45 \times 460.65}{43560}) = \boxed{0.96\ Acres}$

Example 4-3

A parcel of land which adjoins a body of water was surveyed by the open traverse method. The measured distances and bearings are recorded on the plan shown below. The property line data for one side of the parcel is missing.

REQUIRED

(A) What is the length and bearing of side EA ?

(B) What is the total area included within the boundaries ABCDEA shown ?

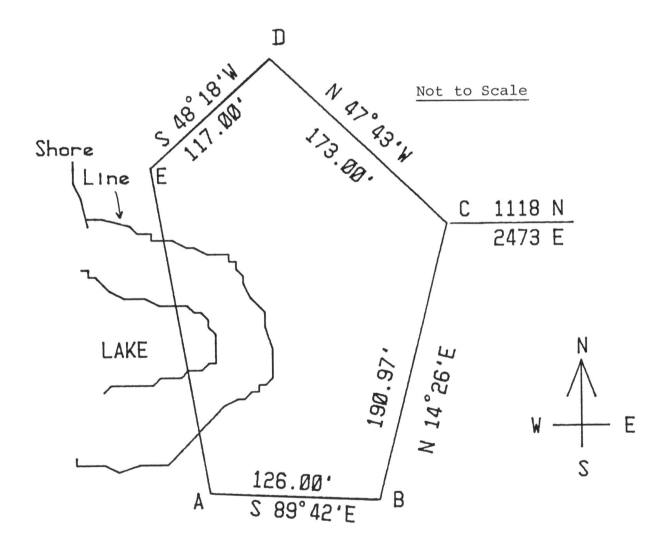

Solutions:

(A) *Using point A as the origin (0,0),*
    calculate the coordinates as follows:

L.	Bearing	Dista.	Latitu. (cos)	Depart. (sin)	P.	Y	X
					A	0.0	0.0
AB	S 89^{0}42'E	126.00	-0.66	126.00			
					B	-0.66	126.00
BC	N 14^{0}26'E	190.97	184.94	47.60			
					C	184.28	173.60
CD	N 47^{0}43'W	173.00	116.39	-127.99			
					D	300.67	45.61
DE	S 48^{0}18'W	117.00	-77.83	-87.36			
					E	222.84	-41.75
					A	0.0	0.0

$$\text{Length } \overline{EA} = \sqrt{222.84^2 + 41.75^2} = \boxed{226.72 \text{ ft}}$$

$$\text{Bearing } \overline{EA} = \tan^{-1} \frac{41.75}{222.84} = \boxed{S \ 10^0 36'42''E}$$

(B) *Calculate the area ABCDEA by the coordinates method.*

$$\text{Area} = \frac{1}{2} \left(\frac{1}{43560}\right) \text{X}$$

$$\left[\begin{array}{l} - \ (-0.66)(173.60) \ - \ (184.28)(45.61) \\ - \ (300.67)(-41.75) \ + \ (126.0)(184.28) \\ + \ (173.6)(300.67) \ + \ (45.61)(222.84) \end{array} \right]$$

$$= \frac{1}{2} \left(\frac{89841.86}{43560}\right)$$

$$= \boxed{1.03 \text{ acres}}$$

Trapezoidal Rule

The area with an irregular boundary can be calculated by the trapezoidal rule. It applies only to the area where the offsets are at regular intervals.

$$\text{Area} = W \left(\frac{h_1}{2} + h_2 + h_3 + \cdots + h_{n-1} + \frac{h_n}{2} \right)$$

where W = offsets spacing
      n = number of offsets

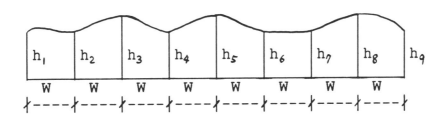

Simpson's one-third Rule

Simpson's one-third rule applies only to the area where the offsets are at regular intervals and has an odd number of offsets. It is a more accurate method for an irregular curved boundary.

$$\text{Area} = \frac{W}{3} \left[h_1 + 2 \times (h_3 + h_5 + \cdots + h_{n-2}) \right.$$
$$\left. + 4 \times (h_2 + h_4 + \cdots + h_{n-1}) + h_n \right]$$

where h_1 = the first offset
      h_n = the last odd-numbered offset

Sample Problems 4

1. Determine the area ABCD (ft^2) in the following figure by the trapezoidal rule.

   (A) 1223   (B) 1332   (C) 1262   (D) 1302

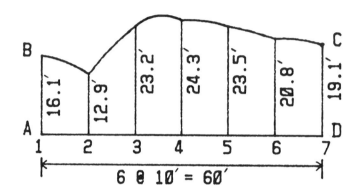

2. Determine the area ABCD (ft^2) in the above figure by the Simpson's one-third rule.

   (F) 1223   (G) 1202   (H) 1262   (J) 1302

3. The coordinates of points A, B, C are shown below. Determine the area inside the traverse ABC (acres) by the coordinates method.

Points	Y	X
A	2,250,000.000	164,500.000
B	2,249,585.478	164,123.475
C	2,249,350.149	164,382.551

   (A) 2.01   (B) 2.52   (C) 2.25   (D) 2.81

ANS :   1.A   2.G   3.C

5. TOPOGRAPHIC SURVEY

Topographic survey is the determination of the location of objects on the surface of the earth. It may be done by aerial or ground method. The first requirement of topographic survey is a good control survey.

The Federal Geodetic Control Committee (FGCC) establishes the following orders of control accuracy in 1974.

Table 5-1   Horizontal Control Accuracy

Order	Closure in length	Principal uses
First-order	1 part in 100,000	National Control Netwok (Primary Control)
Second-order Class I	1 part in 50,000	High-value land areas (Secondary Control)
Class II	1 part in 20,000	Lightly developed areas (Supplemental Control)
Third-order Class I	1 part in 10,000	Local construction
Class II	1 part in 5,000	(Local Control)

Table 5-2   Vertical Control Accuracy

Order	Maximum Closure	Principal uses
First-order Class I	3 mm $\sqrt{K}$	Basic Net A
Class II	4 mm $\sqrt{K}$	Basic Net B (Basic Framework)
Second-order Class I	6 mm $\sqrt{K}$	Metropolitan areas (Secondary Network)
Class II	8 mm $\sqrt{K}$	Local Project (General Area Control)
Third-order	12 mm $\sqrt{K}$	(Local Control)

where K is the distance between Bench Marks in Kilometers

There are two types of control surveys: Horizontal and Vertical. Horizontal control may be done by triangulation, traverse, or trilateration survey. For small areas, traverses furnish satisfactory control. For large areas, such as over a state, triangulation or trilateration furnishes the most economical horizontal control. Vertical control may be done by barometric, trigonometric, or differential leveling. The required accuracy for a control survey depends on its purpose.

Cartography is the technique of making maps. Map scales are generally classified as large, medium, and small.

Large scale: 1 in = 100 ft (1:1,200) or larger.
Medium scale: 1 in = 100 ft to 1000 ft (1:1,200 to 1:12,000)
Small scale: 1 in = 1000 ft (1:12,000) or smaller.

CONTOURS   A contour is a line connecting points of same elevation. For National Standards of Map Accuracy, elevations should be able to be interpolated from a map to within one-half the contour interval. Thus, for an accuracy of 10 ft, a 20-ft contour interval is necessary. Contours are plotted only for elevations evenly divisible by the contour interval. Thus, for a 10-ft interval, elevations of 100, 110, 120, and 130 are shown, but 105, 115 and 125 are not. To improve legibility of contours on the map, every fifth contour line is made heavier (also called INDEX Contour). For example, for a 10-ft interval, the 100,150, and 200 lines are made heavier. The dashed lines may indicate supplementary contours at one-half or one-fourth the basic interval. Sometimes, dashed line represents drainage line.

For average terrain, the following map scales and contour interval relationships generally provide suitable spacing:

English System		Metric System	
Scale (ft/in)	Interval(ft)	Scale	Interval(m)
50	1	1:500	0.5
100	2	1:1000	1.0
200	5	1:2000	2.0
500	10	1:5000	5.0
1000	20	1:10000	10.0

Certain properties of contours are listed as follows:

1. Every contour must close upon itself.
2. Contours are perpendicular to the direction of maximum slope.
3. The distance between contours shows the steepness of a slope.
4. Evenly spaced parallel contours show a uniform slope.
5. Irregular contours show rough, rugged ground.
6. Depressions are often indicated by hachures inside the
   lowest contour.

Hachures are a series of short lines drawn in the direction
of the slope. From the elevations of the contour lines shown
in figure (A), a depression is represented in the figure.
Figure (C) is incorrect, a contour line must close upon itself, or end at the edges of the map. Figure (D) is incorrect,
contour lines can not cross or meet.

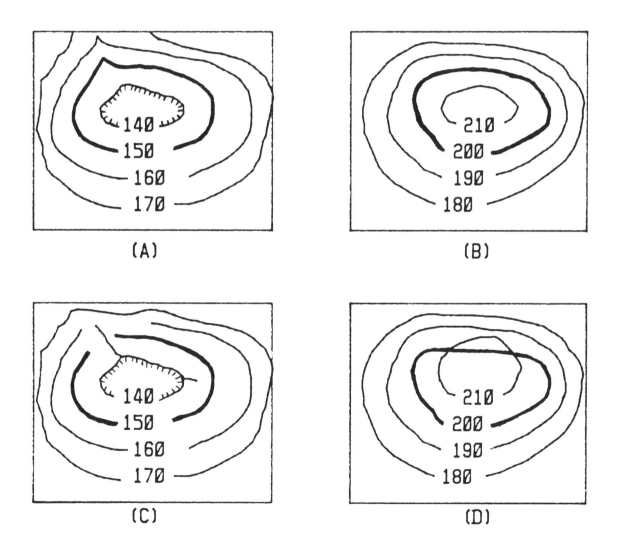

Methods of locating contours:

1. Direct method

   Trace-contour method:
   The rodman selects trial points until the required rod
   reading is obtained. After the point has been located
   by trial, the distance and azimuth are read, and then
   the process repeated. The trace-contour method is the
   most accurate and time consuming method.

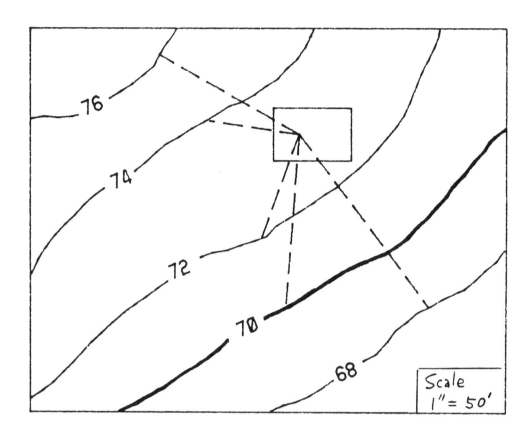

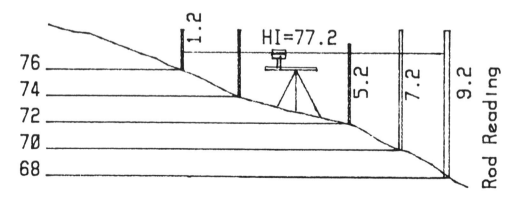

2. Indirect method

   A. Controlling-point Method:
      The rod is set on selected points where there are
      changes in ground slope. Contour lines are determined
      by interpolation. The accuracy of the map depends on
      the experience and judgment of the topographer. This
      method is suitable for large area to a relatively
      small scale.

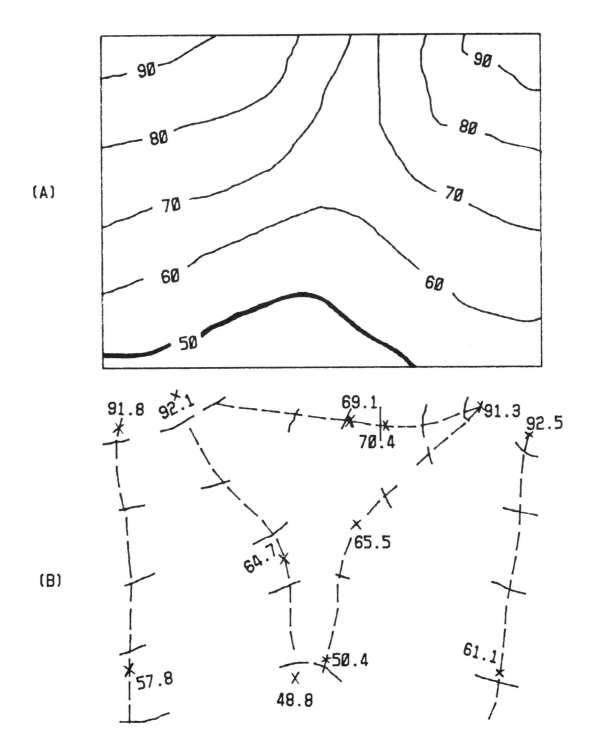

B. Cross-section Method:
   The rod is set on selected stations or plus stations.
   This method is suitable for the preparation of strip
   maps.

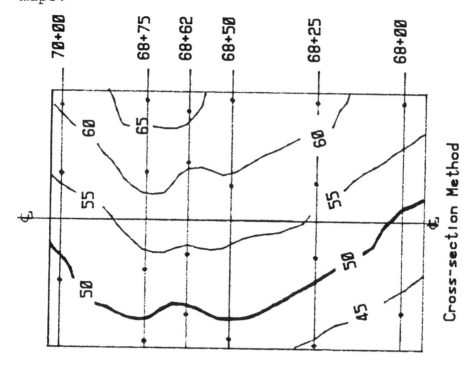

C. Grid Method
   This method is used in the areas of limited extent.
   The area is usually divided into squares or rectangles
   of 25 to 100 ft. It is shown in the following figure.

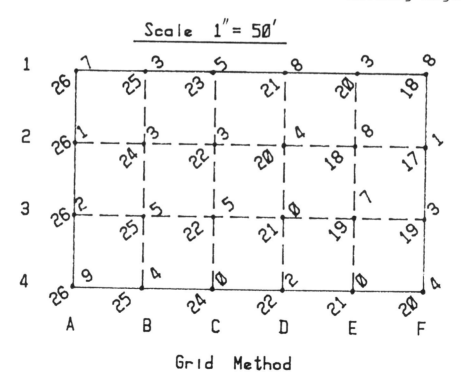

Grid Method

Electronic Distance Measuring (EDM) is a modern method of precise and rapid measurement of slope distances. It has maximum ranges varying from 0.3 to 40 miles. Electronic distance instruments are generally expensive and not very useful for the range of distances used by optical measurement; therefore, tapes, EDMs and optical instruments complement one another.

The Global Positioning System (GPS) is a worldwide system of navigation satellites currently being implemented by the U.S. department of Defense. At the National Geodetic Survey (NGS), the observations of GPS satellite signals yield very accurate relative position data (three-dimensional). The GPS satellite surveying technology has been adopted by the NGS as the primary way to establish geodetic control.

Sample Problems 5

1. The principal uses of the Second Order, Class II, horizontal control survey is to serve

   (A) National Control Network.   (B) High-value land areas.

   (C) Lightly developed areas.   (D) Local construction.

2. The accuracy required for the closure in length of a Second Order, Class I, horizontal control survey is

   (F) 1:100,000   (G) 1:20,000

   (H) 1:50,000   (J) 1:10,000

3. Horizontal control survey may be done by

   I   Triangulation
   II   Traverse
   III   Trilateration

   (A) I and II only   (B) I and III only

   (C) II and III only   (D) I, II and III

4. You are the Project Engineer in charge of a pipeline construc-tion project from Sacramento, California to Reno, Nevada. Which of the following orders of accuracy do you recommend for horizontal control survey for this project?

   (F) First Order   (G) Second Order, Class I

   (H) Third Order   (J) Second Order, Class II

5. Which of the following scales is classified as medium map scale ?

   (A) 1" = 200'   (B) 1" = 1200'

   (C) 1" = 50'   (D) All of the above

6. A control survey is requested for a 55-acre flat area. It is in the countryside, and most of the area is covered by grass. Which of the following methods do you recommend for this control survey?

   (F) Leveling   (G) E.D.M.

   (H) G.P.S.   (J) Chainning

7. For locating contours, the most accurate and time-
   consuming method is

   (A) cross-section method   (B) trace-contour method

   (C) controlling-point method   (D) grid method

8. Which of the following statements is incorrect ?

   (F) Every contour must close upon itself.

   (G) Evenly spaced parallel contours show a uniform slope.

   (H) Contours are perpendicular to the direction of
       maximum slope.

   (J) None of the above.

9. A contour map is needed for a project. The project's area
   has 4% slope (in general). Which of the following map scales
   and contour intervals do you recommend for this project ?

   (A) 1" = 20',   2 ft interval

   (B) 1" = 20',   10 ft interval

   (C) 1" = 50',   1 ft interval

   (D) 1" = 50',   5 ft interval

   In the above question, if the area has 40% slope in general,
   which of the following map scales and contour intervals
   do you recommend for this project ?

   (F) 1" = 20',   2 ft interval

   (G) 1" = 20',   10 ft interval

   (H) 1" = 50',   1 ft interval

   (J) 1" = 50',   40 ft interval

5-9

10. In the contour map, which of the following objects may be represented by the dashed lines ?

    I   Dirt Roads
    II   Storm Sewers
    III Houses

(A) I and II only   (B) II and III only

(C) I and III only   (C) I, II and III

11. There is a construction project in the forest area. An 11-mile long route survey is requested by the client, and you are the Project Engineer in charge of this work. Which of the following maps will you request before this survey?

(F) Aerial photograph   (G) Topographic map

(H) Parcel map   (J) N.G.S. map

Which of the following method is suitable for this route survey?

(A) Photogrammetry   (B) Stadia method

(C) Plane survey   (D) E.D.M.

12. A contour map is requested for an 1 X 1.8 miles rectangular area. It needs to be drawn on a 24" X 36" paper. Which of the following scales is appropriate for this map ?

(A) 1" = 100'   (B) 1" = 200'

(C) 1" = 300'   (D) 1" = 600'

ANS : 1.C   2.H   3.D   4.G   5.A   6.G
        7.B   8.J   9.C F 10.A   11.G C 12.C

6. PHOTOGRAMMETRY

Scale of a vertical photograph is the ratio of photo distance to ground distance. From Fig.6-A, for similar triangles, the scale equation at point A is expressed as:

$$\frac{\overline{ab}}{\overline{AB}} = S = \frac{f}{H - h_A}$$

f : focal length
H : flight elevation
h_A: point A elevation

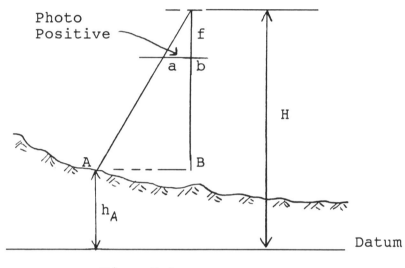

Fig. 6-A

In general, the scale S at any point may be expressed as:

$$\text{Scale} \quad S = \frac{f}{H - h} \quad \text{(Eq.6-1)}$$

f : focal length
H : flight elevation
h : terrain elevation

The C-factor is the ratio of flight height above ground to the contour interval.

$$C = \frac{\text{flight height}}{\text{contour interval}} \quad \text{(Eq.6-2)}$$

Photo scale can be determined from a map scale for the same area. It is calculated as follows:

$$\frac{\text{photo scale}}{\text{map scale}} = \frac{\text{photo distance}}{\text{map distance}} \quad \text{(Eq.6-3)}$$

Example 6-1
_____

A camera for aerial photography has 6 in focal length.
At a flying height 10,000 ft above mean sea level, what
is the photo scale on the ground 2,000 ft above mean sea
level?

Solution:
========

From Eq.(6-1)

$$S = \frac{f}{H - h} = \frac{6 \text{ in}}{(10,000 - 2,000) \text{ ft}} = \frac{6/12}{8,000} = 1:16,000$$

Example 6-2
_____

A plotter is used for mapping and the C-factor is 1200,
at what height should the plane fly above mean terrain in
order to plot 5 ft interval contour?

Solution:
========

From Eq.(6-2)

Flight height = C×(contour interval) = 1,200×(5)
             = 6,000 ft above mean terrain

Example 6-3
_____

On a vertical photograph, the length of an airport runway
is 3.6 in. On a map of scale 1:12,000 , it measures 4.5 in.
What is the photo scale at the runway elevation?

Solution:
========

From Eq.(6-3)

$$\text{photo scale} = \frac{\text{photo distance}}{\text{map distance}} \text{(map scale)} = \frac{3.6}{4.5} \left(\frac{1}{12,000}\right)$$

= 1:15,000   or   1 inch = 1,250 feet

Example 6-4
_____

The scale of the photograph is 1:12,000 ; The camera has
a 6-in focal length with 9X9 in format. Forward overlap is
60%, side lap is 30%. The flight plan is to cover 5 miles
wide and 20 miles long.

REQUIRED
_____

(A) What is the flying height above mean terrain?

(B) What is the distance between exposures?

(C) How many photographs are required per flight line?

(D) What is the distance between flight lines?

(E) How many flight lines are required?

(F) How many photographs will be taken?

(G) How many picture points are needed for photo control?

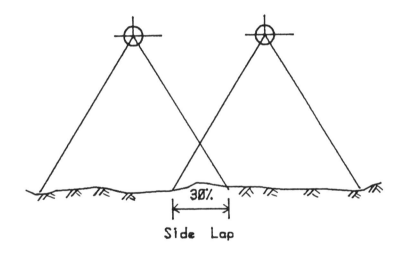

Side Lap

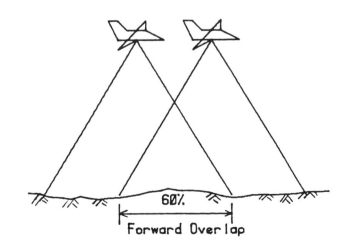

Forward Overlap

Solutions:

(A) Flying height, from Eq.(6-1)

$$S = \frac{f}{H} = \frac{6/12}{H} = \frac{1}{12,000} \quad ; \quad H = 6,000 \text{ ft}$$

(B) $1:12,000 \implies 1 \text{ in} = 1,000 \text{ ft}$

Distance between exposures: $(1 - 0.6) \times (9) \times (1,000) = 3,600 \text{ ft}$

(C) Length of each flight line = 20 miles $\times$ (5,280 ft/mile)

$$= 105,600 \text{ ft}$$

Number of photographs per flight line: $\dfrac{105,600 \text{ ft}}{3,600 \text{ ft/photo}} = 29.3$

(say 30)

Add 1 photo to ensure the coverage, total = 30 + 1 = 31 photos

(D) Distance between flight lines: $(1 - 0.3) \times (9) \times (1,000) = 6,300 \text{ ft}$

(E) Width of the area = 5 miles $\times$ (5,280 ft/mile) = 26,400 ft

Number of lines = $\dfrac{26,400 \text{ ft}}{6,300 \text{ ft/line}} = 4.19$ (say 5)

Add 1 line to ensure the coverage, total = 5 + 1 = 6 lines

(F) Total number of photos required:

Total photos = (31 photos/line) $\times$ (6 lines) = 186 photos

(G) A network of picture points of known position is used
as a reference of photo control to fix the detail of
aerial photographs. The density and distribution of
field control points to be photo-identified (called
picture points) are determined by the desired accuracy,
the characteristics of the photography, and the type
of photogrammetric equipment.

7. CONSTRUCTION SURVEY

STAKES for the highway construction are usually set at 100 or 50 feet intervals on tangents at uniform grade. Stakes are set at 50 or 25 feet intervals on horizontal or vertical curves. Stakes for pavement must be set on an offset line (usually 2 ft) from the edge of pavement; thus, stakes will not be destroyed by construction equipment.

GRADE STAKES are driven at points having the same ground and grade elevation. Fig.7-A, three sections occur in passing from fill to cut, and a grade stake is set for each section.

SLOPE STAKES are driven at the intersection of the ground and a planned slope. The cut 'C' or fill 'F' is marked on the slope stake. Actually, there is no cut or fill at the location of the slope stake. Fig.7-A, the C or F given is the vertical distance from slope stake elevation to grade. The slope is expressed as a ratio of horizontal to vertical distance. Thus, 2:1 slope means a 1 ft vertical change for each 2 ft horizontal distance. The stake marked ' C 2.1 @ 34.2 ' means that the stake is 34.2 feet from the center line and 2.1 feet above grade. The stake marked ' F 3.4 @ 36.8 ' means that the stake is 36.8 feet from the center line and 3.4 feet below grade (finish elevation).

GRADE POINT: a fill (or cut) section meets the natural ground.

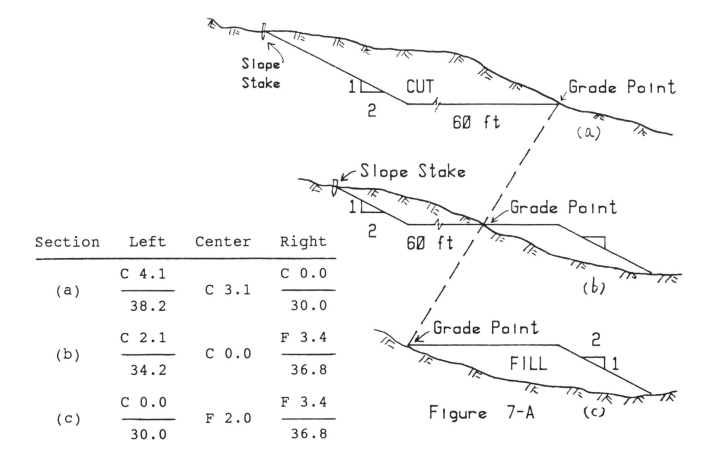

Section	Left	Center	Right
(a)	C 4.1 ——— 38.2	C 3.1	C 0.0 ——— 30.0
(b)	C 2.1 ——— 34.2	C 0.0	F 3.4 ——— 36.8
(c)	C 0.0 ——— 30.0	F 2.0	F 3.4 ——— 36.8

Figure 7-A

BLUE TOPS: the tops of the stakes are marked with blue keel and the stakes are driven to grade elevation. The blue tops are normally driven on center line. A guard stake is driven at an angle beside the blue top. The guard stake is marked 'G' to indicate that the stake is driven to grade.

GRADE ROD: the difference between H.I. (Height of Instrument) and grade elevation (finish elevation).

GROUND ROD: a rod reading on the ground.

PRECISION For earthwork, elevations are set to the nearest 0.1 ft.
          For points on the structure (concrete, pipelines), elevations are set to the nearest 0.01 ft.

The most common method for calculating earthwork volume is the Average End Area Method. It is expressed as follows:

$$V = \frac{L\,(A_1 + A_2)}{2 \times 27} = \frac{L}{54}\,(A_1 + A_2)$$

V: Cubic yards

A: Vertical cross section (ft^2).

L: Horizontal distance (ft) between A_1 and A_2.

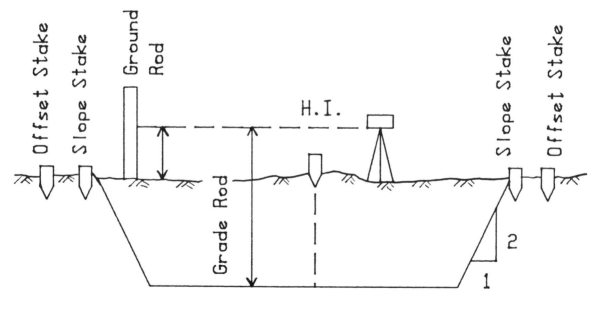

Slope 1:2

For the examples of slope staking, Caltrans Surveys are re-printed on the following pages.

Definitions of slope staking
_____

1. BSR - Begin Slope Rounding, the point on a planned slope where the plane slope changes into either a circular or a parabolic slope.

2. CP - Catch Point, where original ground surface is intersected by a planned plane-slope, either cut or fill, or by a projection of the plane slope.

3. RPSS - Reference Point to Slope Stake.

4. Ground Stake - the RP of an RPSS is a ground stake.

5. HP - Hinge Point, the beginning point of a side slope at or near the edge of a roadbed.

6. SS - Slope Stake, a slope-grading stake which is set at the intersection of the ground and a planned slope.

7. Witness Stake - RP and slope-grading data are written on this stake.

8. CGS - Contour Grading Stake.

9. ESR - End Slope Rounding.

_____

Example 7-1
_____

You have set the slope stake shown in figure A. The elevation of the point set is 216.02. Markings of horizontal and vertical slope references are from the ESR to each of the other grade points. What is the elevation of the ESR, BSR, and HP referred to on the stake?

Solution:
================

The elevation is:

  ESR = 216.02 + 1.5 = 217.52

  BSR = 217.52 - 10.9 = 206.62

  HP = 217.52 - 18.2 = 199.32

Figure A.

7-3

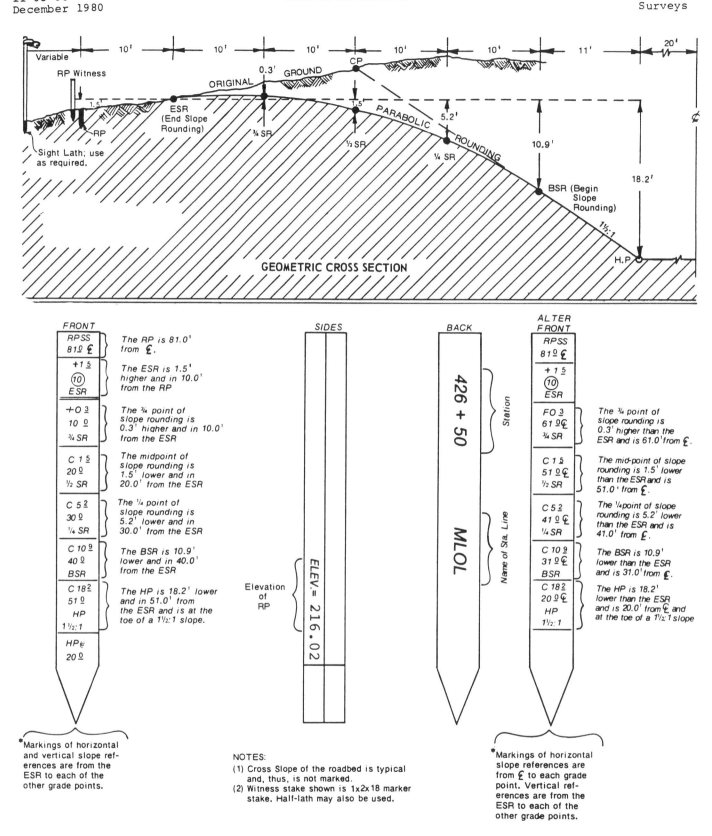

Fig. 11-05-C Slope Staking-Excavation with Parabolic Slope Rounding.

Example 7-2

You have set the slope stake shown in figure B.
The elevation of the point set is 882.05.
Markings of horizontal and vertical slope
references are from a grade point to the
next grade point. What is the elevation
of the CP, Gutter (Gtr.), and HP referred
to on the stake?

Solution:

The elevation is:

CP = 882.05 - 1.0 = 881.05

Gutter = 881.05 - 10.0 = 871.05

HP = 871.05 + 1.0 = 872.05

Figure B

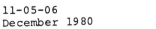

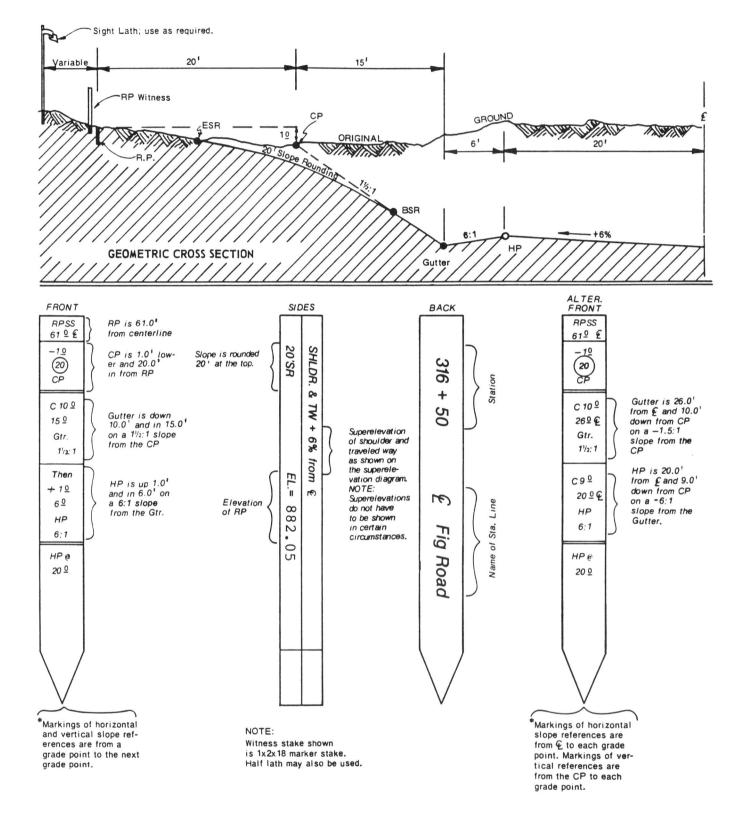

MARKINGS on RP WITNESS STAKE*

Fig. 11-05-B Slope Staking - Excavation

Sample Problems 7

1. You are the resident engineer of a highway construction project. The construction stakes for earthwork are usually set to the nearest

    (A) 0.05 ft   (B) 0.1 ft   (C) 0.5 ft   (D) 1.0 ft

2. The stake marked ' C 4.1 @ 38.2 ' means that the stake is

    (F) 38.2 feet from the center line and 4.1 feet above grade.

    (G) 38.2 feet from the center line and 4.1 feet below grade.

    (H) 38.2 feet from the edge line and 4.1 feet above grade.

    (J) 38.2 feet from the edge line and 4.1 feet below grade.

3. Blue Top is a construction stake

    (A) driven to grade elevation.

    (B) driven at point having the same ground and grade elevation.

    (C) driven at the intersection of the ground and a planned slope.

    (D) None of the above.

4. You are a project engineer in charge of highway construction. The stakes on tangent lines at uniform grade are usually set at

    (F) 25' or 50' intervals.   (G) 200' or 100' intervals.

    (H) 50' or 100' intervals.   (J) 25' or 50' or 100' intervals.

5. In the construction field, the stakes are set on an offset line from the edge of pavement. The purpose of setting these stakes is to

    (A) set control level.   (B) set control point.

    (C) avoid being destroyed by construction equipment.

    (D) none of the above.

6. A construction work is completed, and a survey is requested to check the work with design plans. This survey is called

    I   Final Survey
    II   As-Built Survey
    III   Plan Survey

    (F) I and III only   (G) I and II only

    (H) II and III only   (J) I, II and III

7. In the construction field, the guard stake is marked 'G' to indicate that the stake is driven to

    (A) grade elevation   (B) ground elevation

    (C) guard stake   (D) ground rod reading

8. You are the Resident Engineer of a building construction project. The construction stakes for concrete slab are usually set to the nearest

    (A) 0.05 ft   (B) 0.001 ft   (C) 0.02 ft   (D) 0.01 ft

9. For earthwork, the side slope marked '2:1' means

(F)
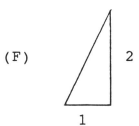

(G)
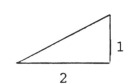

(H)

(J)

10. Which of the following is not the member of a survey team?

    I   Chairman   II   Recorder
    III Instrument Man   IV   Party Chief

    (A) I only   (B) I and II

    (C) I and III   (D) II and IV

11. You have set the slope stake shown in figure W. The elevation of the point set is 94.8. What is the elevation of the toe of slope referred to on the stake?

    (A)   Elev. 92.7   (B) Elev. 96.8

    (C)   Elev. 92.8   (D) Elev. 90.7

12. You have set the slope stake shown in figure X. The elevation of the point set is 82.48. Markings of horizontal and vertical slope references are from a grade point to the next grade point. What is the elevation of the HP?

    (A) 104.48   (B) 106.48   (C) 134.48   (D) 136.48

13. You have set the slope stake shown in figure Y. The elevation of the point set is 66.91. Markings of horizontal and vertical slope references are from CP#1 to each of the other grade points. What is the elevation of CP#2?

    (F) 73.41   (G) 71.41   (H) 69.41   (J) 61.41

What is the elevation of Lt.HP?

    (A) 80.41   (B) 79.41   (C) 75.41   (D) 72.41

14. You have set the Contour-Grading stake shown in figure Z. What is the elevation of the stake?

    (F) 82.1   (G) 81.2   (H) 82.4   (J) 80.2

What is the elevation of the flow line (F.L.)?

    (A) 80.2   (B) 81.2   (C) 81.1   (D) 82.1

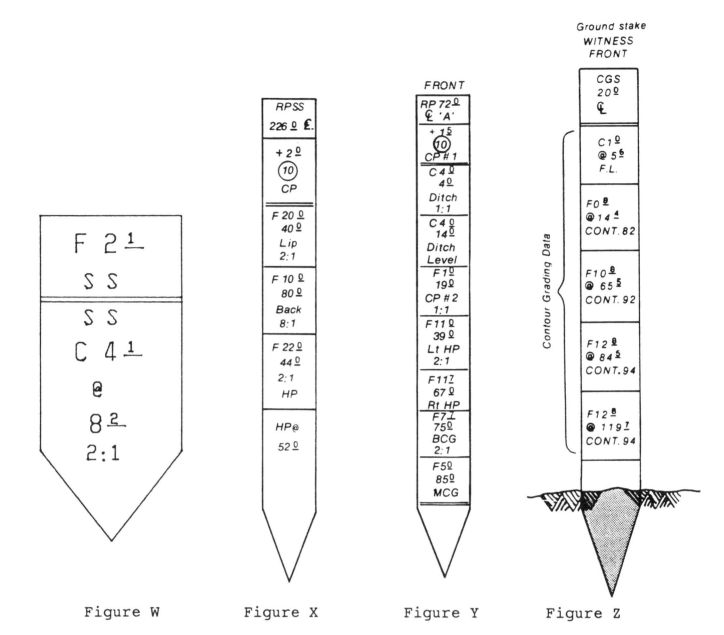

Figure W Figure X Figure Y Figure Z

15. The F.G.S. illustrated below is 'Final-Grade Stake'. The term 'Final-Grade Stake' was adopted as a term to replace 'Blue Top'. The elevation of the P.P.P (Plane of Pavement as Produced) is 1091.17. What is the elevation at the toe of F.G.S.?

(A) 1090.93   (B) 1090.57

(C) 1090.53   (D) 1090.81

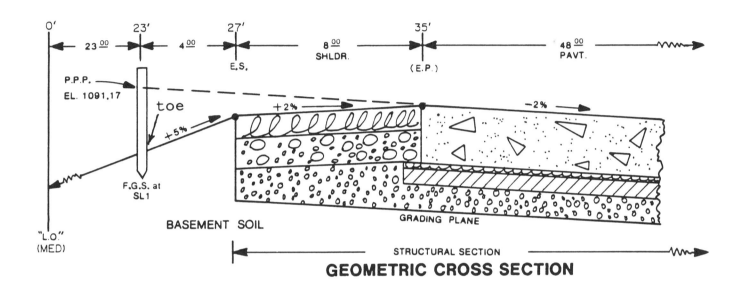

GEOMETRIC CROSS SECTION

ANS : 1.B   2.F   3.A   4.H   5.C
      6.G   7.A   8.D   9.G   10.A
      11.C   12.D   13.H B   14.G A   15.B

8. LEVELING

LEVELING is a process to determine the vertical position of different points below, on, or above the ground.

BENCH MARK is a permanent point of known elevation.

VERTICAL LINE follows the direction of gravity to the center of the earth.

HORIZONTAL LINE is perpendicular to the Vertical Line.

LEVEL SURFACE is a curved surface which has every point perpendicular to the direction of gravity.

LEVEL LINE is a line in a Level Surface, hence is a curved line.

DATUM : Any level surface to which elevations are referred. Mean Sea Level is usually used for a datum.

DIFFERENTIAL LEVELING is the most commonly used leveling method.

An example for differential leveling is illustrated below.

LEVEL   NOTES

STATION	BS (+S)	HI	FS (-S)	Elevation
BM1	7.25			107.34
		114.59		
TP1	9.66		1.07	113.52
		123.18		
TP2	8.09		0.48	122.70
		130.79		
BM2			10.70	120.09

Table 8-1 (See Figure 8-A)

Terminology

BM   Bench Mark
TP   Turning Point
BS   BackSight (also called plus sight)
FS   ForeSight (also called minus sight)
HI   Height of the Instrument

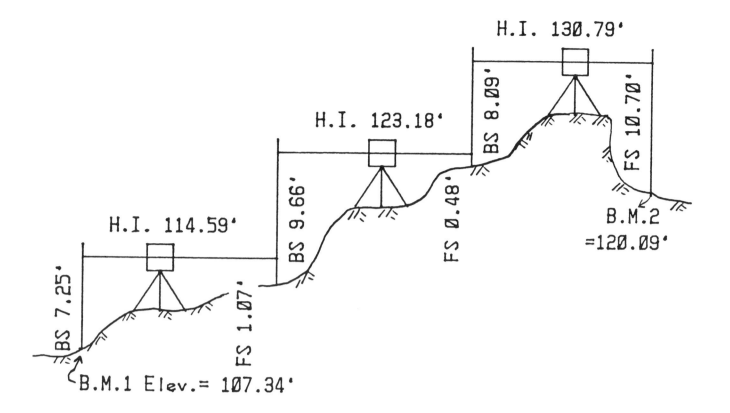

Figure 8-A   Profile view of differential leveling

Height of the Instruments:

  HI1 = 107.34 + 7.25 = 114.59

  HI2 = 114.59 + 9.66 - 1.07 = 123.18

  HI3 = 123.18 - 0.48 + 8.09 = 130.79

Bench Marks:

  BM1 = 107.34

  BM2 = 130.79 - 10.70 = 120.09

  The difference in elevation = 120.09 - 107.34 = 12.75

Check:

  The backsight sum = 7.25 + 9.66 + 8.09 = 25.0

  The foresight sum = 1.07 + 0.48 + 10.7 = 12.25

  (BS sum) - (FS sum) = 25.0 - 12.25 = 12.75 (OK)

STADIA TACHEOMETRY is a commonly known procedure to measure horizontal and vertical distance.

Stadia sighting depends on two hair-lines placed at equal distances above and below the central line in a telescope. A vertical rod is sighted and the distance on the rod between the two stadia hairs is read as intercept. Modern internal-focusing instruments have fixed the distance between the hair-lines so that when the telescope is horizontal and the rod is vertical, distance D equals 100 times the intercept S as in Figure 8-B. (Stadia interval factor K = 100)

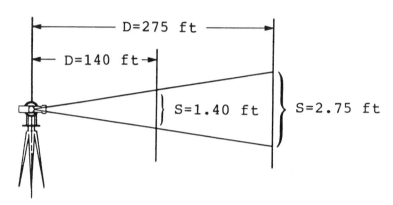

Figure 8-B

If the sighting is inclined, as in Figure 8-C, then

$$D = 100 \; S \cos \alpha$$

$$H = D \cos \alpha = 100 \; S \; (\cos^2 \alpha)$$

$$V = D \sin \alpha = 100 \; S \; (\cos \alpha)(\sin \alpha)$$

$$= \frac{1}{2} \; 100 \; S \; (\sin 2\alpha)$$

where

    S : intercept reading

    α : vertical angle

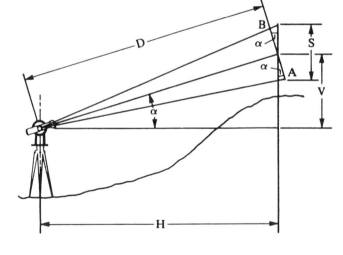

Figure 8-C

The Philadelphia rod is commonly used in leveling. It consists of two sections, the front section is used for reading up to 7 ft, the rear section begins at 7 ft at the top and increases downward to 13 ft. The rear section is to be extended for reading between 7 and 13 ft. The graduations on the face of the rod are in hundredths of a foot. The readings on the rods shown in the figure below are as follows:

(A) 2.636 ft

(B) 10.052 ft

(C) 7.807 ft

(D) s = 3.850 ft

     t = 4.050 ft

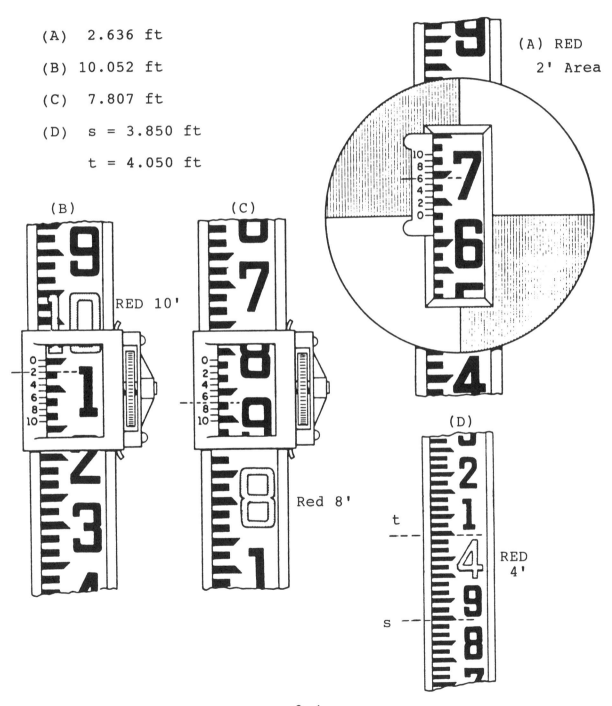

(A) RED 2' Area

RED 10'

Red 8'

(D) RED 4'

Example 8-1

Stadia reading are made from an instrument set up to two different points where the rod was held (Stadia interval factor K = 100). The recorded values are $S_1 = 0.925$ m, $\alpha_1 = 2°15'$; and $S_2 = 1.541$ m, $\alpha_2 = -5°45'$. Determine the horizontal distance between the instrument and the points.

Solution:

$$H_1 = 100 \times 0.925 \cos^2 2°15' = 92.36 \text{ m}$$

$$H_2 = 100 \times 1.541 \cos^2 5°45' = 152.55 \text{ m}$$

Example 8-2

The height of the instrument of the above example is measured as 1.58 m. The central line reading of the telescope for the two observations are $R_1 = 1.71$ m; $R_2 = 1.44$ m. Determine the difference in elevation between the point over which the instrument was set up and the points where the rod was held.

Solution:

$$V_1 = 50 \times 0.925 \sin 4°30' = 3.63 \text{ m}$$

$$V_2 = 50 \times 1.541 \sin (-11°30') = -15.36 \text{ m}$$

Difference in elevation:

$$DV_1 = 1.58 + 3.63 - 1.71 = 3.50 \text{ m}$$

$$DV_2 = 1.58 - 15.36 - 1.44 = -15.22 \text{ m}$$

Sample Problems 8

1. A field survey has been performed for a proposed sewer extension based on two different alignments. The field leveling notes are shown below.

Centerline Profile
Natural Ground

Alignment A

Station	Elevation
0+00	338.1
1+00	336.9
2+00	349.2
3+00	352.3
4+00	353.5
5+00	352.8
6+00	353.1

Alignment B

Station	Elevation
0+00	356.3
1+00	348.2
2+00	339.3
3+00	326.7
4+00	334.8
5+00	351.3
6+00	356.1

The proposed sewer invert will begin at Station 0+00 at an elevation of 322.0 at the existing sewer with a slope of +2%. Which of the statements are not true regarding the two alignments?

    I   Alignment B will require more excavation than A.
   II   Alignment B will not have pipe exposed.
  III   Alignment B has the maximum trench depth compared to A.

   (A) I and II only

   (B) I and III only

   (C) II and III only

   (D) I, II, and III

2. You are given a profile sheet with the beginning station of 1+45, elevation 348.26 and the following:

   (1) slope -0.26 percent to Station 4+25
   (2) slope +1.35 percent to Station 10+35
   (3) slope -10.5 percent to Station 12+50

   What is the elevation at Station 12+50?

   (A) 327.39   (B) 317.47   (C) 333.19   (D) 333.14

3. If, in running a line of levels to establish a bench mark, one backsight is misread as 5.43 instead of 6.43 and one foresight is misread as 6.78 instead of 7.78, the net effect on the elevation of the bench mark established will be which of the following:

   (A) to show it two feet too high

   (B) to show it one foot too low

   (C) to show it two feet too low

   (D) none of the above

4. The following notes were taken during a differential leveling.

Station	BS	FS	Elevation
BM1	3.27		32.45
TP1	4.56	4.31	
TP2	5.34	4.28	
BM2	6.72	5.97	
TP3	4.58	3.25	
TP4	8.21	4.31	
BM3		5.61	

   Determine the elevation of BM2.

   (F) 33.84   (G) 37.78   (H) 34.51   (J) 31.06

   Determine the difference in elevation between BM2 and BM3.

   (A) 6.70   (B) 6.34   (C) 4.95   (D) 3.56

5. The figure below shows a Philadelphia rod face detail. It is shown to 1/100 of a foot. Determine the reading of the rod.

   (F) 3.64

   (G) 3.76

   (H) 3.65

   (J) 3.75

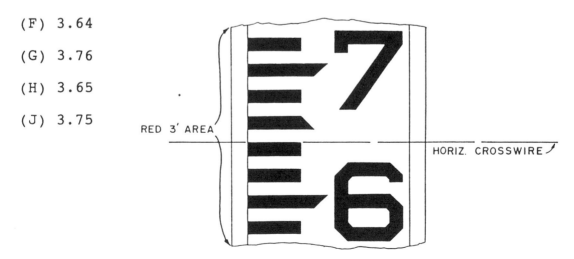

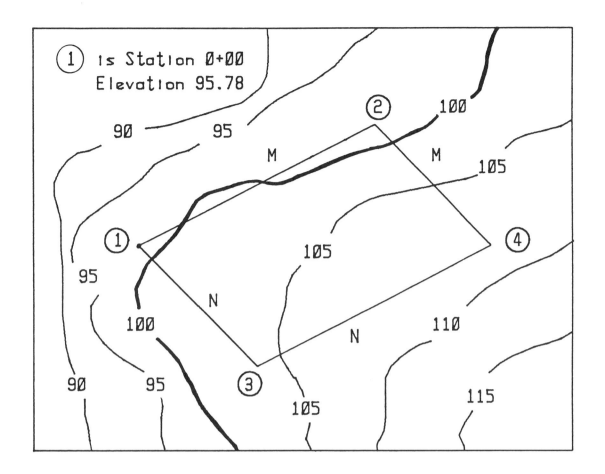

Scale   1"=100'

6. A contour plan has been requested for a proposed sewer
   extension project. There are two different alignments,
   (M):1-2-4, and (N):1-3-4 as shown in the figure above.

   The proposed sewer invert will begin at Station 0+00
   at an elevation of 95.78 ft with a slope of 2%.
   Which of the statements are true regarding the two
   alignments?

     I   Alignment N will require more excavation than M.
    II Alignment M will have pipe exposed.
   III Alignment N has the maximum trench depth compared to M.

   (A) I and II only   (B) I and III only

   (C) II and III only   (D) I, II, and III

       ANS :   1.A   2.C   3.D   4.J B   5.F   6.D

9. ENGINEERING PRACTICE

A Civil Engineer registered prior to January 1, 1982 may practice LAND SURVEYING provided that Civil Engineer is fully competent and proficient to do so. Also, any registered Civil Engineer may practice ENGINEERING SURVEYING, as defined in the following:

(1) Location, relocation, establishment, reestablishment, or retracement of the alignment or elevation for any of the fixed works embraced within the practice of civil engineering.

(2) Determination of the configuration or contour of the earth's surface, or the position of fixed objects thereon or related thereto, by means of measuring lines and angles, and applying the principles of trigonometry or photogrammetry.

United States Public Land Surveys:

Township is the primary unit for the subdivision of public lands. It is approximately 6 miles square and bounded by meridians and parallels of latitude. The township is divided into 36 sections, each approximately 1 mile square.

The township is located with respect to principal axes passing through an origin called an Initial Point. Through this point, the north-south axis is a true meridian called the Principal Meridian, and the east-west axis is a parallel of latitude called the Base Line.

A row of townships extending north-south is called a Range (R), and a row extending east-west is called a Tier (T). For example, 'T2S, R3W, 4th PM' designates a township in the second tier south of the base line and the third range west of the fourth principal meridian. Sometimes, the word 'township' is substituted for 'tier'. Thus, 'T2S, R3W', read 'township two south, range three west'.

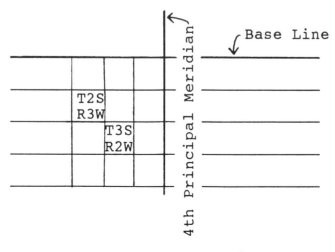

Example 9-1
_____

As a part of a large subdivision development, you are studying
alternate sites for locating a circular steel water tank. One
of the possible sites is indicated in the following sketch.
The site is encumbered by an existing 20-foot wide electrical
easement. The tank must avoid the easement.

The water district has imposed the following additional requirements regarding the tank location. These setbacks are measured
to the face of the tank.

    1. Setback from easement = 0 ft
    2. Setback from boundary lines = 30 ft

The centerline of the west easement can be described as follows:

    Beginning at a point on the northwesterly side of Lot 5 in
    Tract 12345, said point having coordinates (N 2,249,919.43,
    E 164,426.81), thence to a point having coordinates
    (N 2,249,416.60, E 164,309.77)

The coordinates in the above description are based on the
California Coordinate System, Zone II.

REQUIREMENTS
_____

(A) Draw to scale and label the easement, setbacks on Fig.9-A.

(B) Determine the maximum diameter of the water tank that can
    be supported by this site. Record your answer to the
    nearest 0.1 ft.

(C) What is the amount of the encumbered area on the northwest
    side of the electrical easement? Record your answer to the
    nearest 0.1 ft.

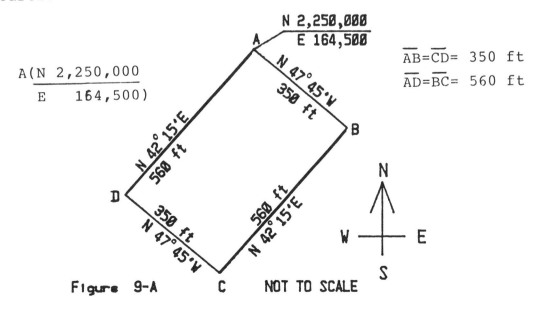

Figure 9-A   NOT TO SCALE

9-2

Solution:

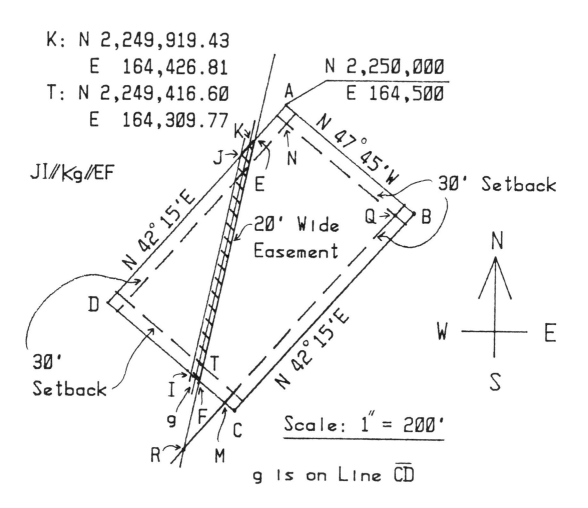

K: N 2,249,919.43
    E 164,426.81
T: N 2,249,416.60
    E 164,309.77

JI∥Kg∥EF

N 2,250,000
E 164,500

30' Setback

20' Wide
Easement

N 42°15'E

30'
Setback

N 47° 45'W

N 42°15'E

Scale: 1" = 200'

g Is on Line $\overline{CD}$

Figure 9-B

Solutions:

(A)
a. *Find point K*

$$x = 164500 - 164426.81 = 73.19'$$

$$y = 2250000 - 2249919.43 = 80.57'$$

$$\text{Length } \overline{KA} = \sqrt{80.57^2 + 73.19^2} = 108.85'$$

$$\text{Bearing } \overline{KA} = \tan^{-1}\left(\frac{73.19}{80.57}\right) = N\ 42^0 15'E$$

Point K is on line $\overline{AD}$.

b. *Find points D,C*

Point D : $2250000 - 560\ (\cos 42.25^0) = N\ 2249585.478$

$164500 - 560\ (\sin 42.25^0) = E\ 164123.475$

Point C : $2249585.478 - 350\ (\cos 42.25^0) = N\ 2249350.149$

$164123.475 - 250\ (\sin 42.25^0) = E\ 164382.551$

c. *Find point T*

$$x = 164382.551 - 164309.77 = 72.78'$$

$$y = 2249416.60 - 2249350.149 = 66.45'$$

$$\text{Length } \overline{CT} = \sqrt{66.45^2 + 72.78^2} = 98.552'$$

$$\text{Bearing } \overline{CT} = \tan^{-1}\left(\frac{72.78}{66.45}\right) = N\ 47^0 36.2'W \neq N\ 47^0 45'W$$

$$\text{Angle difference} = 47^0 45' - 47^0 36.2' = 0.147^0$$

Point T is not on line $\overline{CD}$.

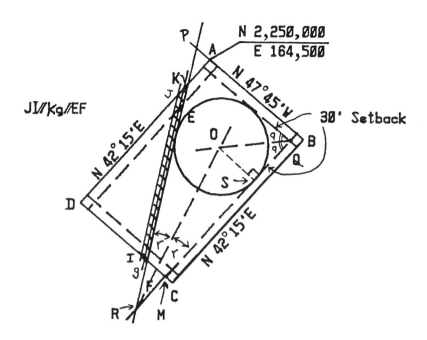

JI∥Kg∥EF

(B) The maximum circle inside a triangle is shown above.

Diameter $\overline{OS} = \overline{RO} \; (\sin r)$

$$r = \frac{1}{2} \; (\angle ERQ) \qquad\qquad q = \frac{1}{2} \; (\angle PQR) = 45^{0}$$

a. Find r

Bearing $\overline{RE} = \tan^{-1} \left(\dfrac{164426.81 \; - \; 164309.77}{2249919.43 \; - \; 2249416.6} \right)$

$$= N \; 13^{0}06'11''E \quad (= 13.103^{0})$$

$$\angle ERQ = 42.25^{0} - 13.103^{0} = 29.147^{0}$$

$$r = \frac{1}{2} \; (\angle ERQ) = \frac{1}{2} \; (\; 29.147^{0}) = 14.574^{0}$$

b. Find $\overline{RO}$

In $\triangle$ ROQ, $\qquad \dfrac{\overline{RQ}}{\sin \; (\angle ROQ)} = \dfrac{\overline{RO}}{\sin \; q}$

and $\overline{RQ} = \overline{RM} + \overline{MQ}$

$\overline{RM} = \overline{MF} / \tan \; (\angle ERQ) \qquad ; \qquad \overline{MF} = \overline{Cg} - 30 - \overline{gF}$

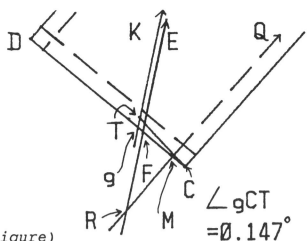

c. Find $\overline{Cg}$ (See the above figure)

$\angle KgC = \angle ERQ + 90^0 = 119.147^0$

$\angle gTC = 180^0 - 119.147^0 - 0.147^0 = 60.706^0$

In $\triangle TgC$, $\dfrac{\overline{CT}}{sin\ 119.147^0} = \dfrac{\overline{Cg}}{sin\ 60.706^0}$

From (A), c. $\overline{CT} = 98.552'$; therefore $\overline{Cg} = 98.411'$

d. Find $\overline{gF}$ (See the figure below)

$\alpha = \angle ERQ = 29.147^0$

$\overline{gF} = 10'/(cos\ \alpha) = 11.45'$

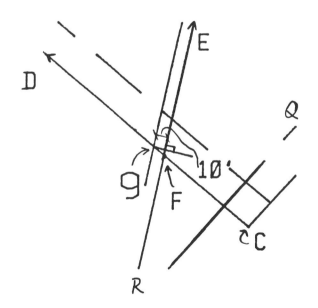

e. Find $\overline{RQ}$ (See Fig.9-B)

$\overline{MF} = \overline{Cg} - 30 - \overline{gF} = 98.411 - 30 - 11.45 = 56.961'$

$\overline{RM} = \overline{MF}/(tan\ 29.147^0) = 102.142'$

$\overline{RQ} = \overline{RM} + \overline{MQ} = 102.142 + 560 - 30 = 632.142'$

f. Find the maximum diameter of the water tank.

In $\triangle$ ROQ, $\dfrac{\overline{RQ}}{sin\ (\angle\ ROQ)} = \dfrac{\overline{RO}}{sin\ q}$

From e. $\overline{RQ}$ = 632.142'; $\angle$ ROQ = $180^0 - r - q = 120.426^0$

Therefore, $\overline{RO}$ = 518.381'

Diameter $\overline{OS} = \overline{RO}\ (sin\ r)$

$= 518.381\ (sin\ 14.574^0)$

$=$ | 130.44 ft |

(C) Encumbered Area IJD

a. $\overline{JD} = \overline{AD} - \overline{AK} - \overline{KE}$

From (A), a. $\overline{AK}$ = 108.85'

Find $\overline{KE}$

$\alpha = \angle$ ERQ = 29.147^0

$\overline{KE}$ = 10'/(sin α) = 20.532'

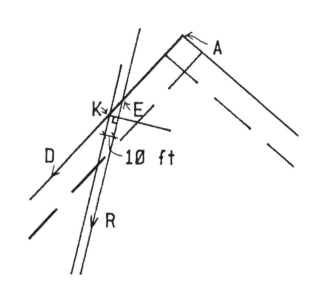

$\overline{JD} = \overline{AD} - \overline{AK} - \overline{KE}$ = 560 - 108.85 - 20.532 = 430.618'

b. $\overline{DI} = \overline{DC} - \overline{Cg} - \overline{gF}$

From (B), c. $\overline{Cg}$ = 98.411'

From (B), d. $\overline{gF}$ = 11.45'

$\overline{DI} = \overline{DC} - \overline{Cg} - \overline{gF}$ = 350 - 98.411 - 11.45 = 240.139'

Area IJD = $\dfrac{1}{2}\ (\overline{JD})\ (\overline{DI})$ = | 51704.1 ft^2 |

Sample Problems 9

1. Which of the following types of surveys can be performed by a civil engineer who was registered after 1982 ?

      I   Boundary surveys   II Topographic surveys
   III   Construction staking surveys

   (A) I and II only   (B) I and III only
   (C) II and III only   (D) I, II, and III

2. Upon passing your RCE test, you open a business by yourself. A client asks you to perform the following tasks:

   (1) Stake for Grading   (2) Property Survey
   (3) Legal Descriptions   (4) Topographic Survey
   (5) Determine the area of a property given boundary limits
   (6) Stake for building construction providing control is established by a person authorized to perform Land Surveying.
   (7) ALTA (American Land Title Association) survey on completed project.
   (8) Design Grading Plan providing no boundary is shown.
   (9) Parcel map

Which of the above can be legally performed by you as a registered civil engineer ?

   (A) All of the above   (B) 1,4,5,6,8
   (C) 1,4,5,6,7   (D) 1,4,5,8,9

3. A township is in the third tier south of the base line and the second range west of the principal meridian. This township can be named as

   (A) T3S, R2W   (B) 3TS, 2RW

   (C) T3SB, R2WP   (D) TS3, RW2

     ANS : 1.C   2.B   3.A

ANSWERS to Sample Problems

Sample Problems 1

Problem A

1. $I = 22°37' + 18°14' = 40°51'$

2. $T = R \tan (I/2)$

    $T = (PC \text{ to } A) + (A \text{ to } PI)$

    $(PC \text{ to } A) = (69+75) - (65+13) = 462'$

    $(A \text{ to } PI) = M \quad ; \qquad \dfrac{614.75'}{\sin 139°09'} = \dfrac{M}{\sin 18°14'}$

    $M = 294.07' \quad ; \qquad T = 462 + 294.07 = 756.07'$

    $R = \dfrac{756.07'}{\tan (40°51'/2)} = 2030.30'$

3. $L = R \triangle = 2030.30 \times (\dfrac{40°51'}{180°}) \pi = 1447.54'$

4. $PT = (65+13) + (14+47.54) = 79+60.54$

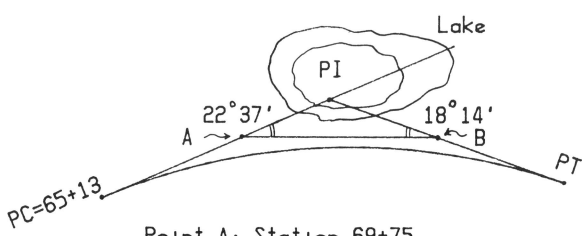

Point A: Station 69+75

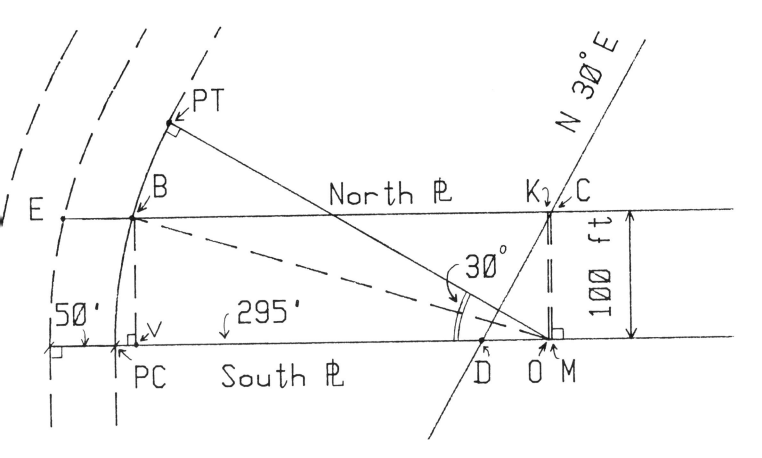

Problem B

5. $\overline{OB} = 400 - 50 = 350'$; $\overline{BV} = 100'$

$$\sin \theta = \frac{\overline{BV}}{\overline{OB}} = \frac{100'}{350'} \qquad ; \qquad \theta = 16.6°$$

6. PC–B $= 350 \times (16.6°/180°)\pi = 101.4'$

7. $\overline{BC} = \overline{BK} + \overline{KC}$; $\overline{BK} = \overline{VO}$; $\tan 16.6° = \dfrac{100}{\overline{VO}}$

   $\overline{BK} = \overline{VO} = 335.44'$

   $\overline{KC} = \overline{DM} - \overline{DO}$; $\overline{DO} = 350 - 295 = 55'$

   $\overline{DM} = 100 \tan 30° = 57.74'$; $\overline{KC} = 57.74 - 55 = 2.74'$

   $\overline{BC} = \overline{BK} + \overline{KC} = 335.44 + 2.74 = 338.2'$

8. $\cos 30° = \dfrac{100}{\overline{CD}}$; $\overline{CD} = 115.5'$

9. $\overline{EB} = \overline{EK} - \overline{BK}$; $\overline{EK} = \sqrt{400^2 - 100^2} = 387.3'$

   $\overline{BK} = 335.44'$; $\overline{EB} = 387.3 - 335.44 = 51.9'$

A–2

Problem C

10. $\sqrt{1000^2 - 160^2} = 987.117'$
    b = 150 - (1000 - 987.117) = 137.12'

11. $\sqrt{1000^2 - 80^2} = 996.8'$
    a = 150 - (1000 - 996.8) = 146.8'

Problem D

12. X = 194.02 - 170 (sin 70°10') - 100.5 (sin 19°50') = 0.
     Y = 36.86 + 170 (cos 70°10') - 100.5 (cos 19°50') = 0.

13. $\overline{OA} = \sqrt{62.71^2 + 247.65^2} = 255.47'$

14. $\tan^{-1} \dfrac{62.71}{247.65} = 14.2098° = 14°12'35.2''$

     ∠ BAO = 5°10'00" + (14°12'35") = 19°22'35"

15. $\dfrac{100.5}{\sin (19°22'35'')} = \dfrac{255.47}{\sin (\angle ABO)}$

     ∠ ABO = 57.497° = 57°29'49" (N.G.)
     From the figure below, ∠ ABO > 90°;
     ∠ ABO = 180° - 57°29'49" = 122°30'11"

16. ∠ BOA = 180° - (122°30'11") - (19°22'35") = 38°07'14"
    ∠ BOC = ∠ BOA + 14°12'35" + 19°50'00" = 72°09'49"
    Arc length BC = 100.5 (72°09'49"/180)π = 126.58'

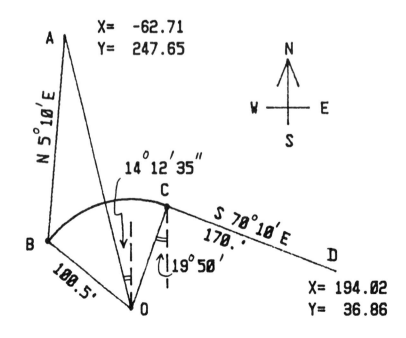

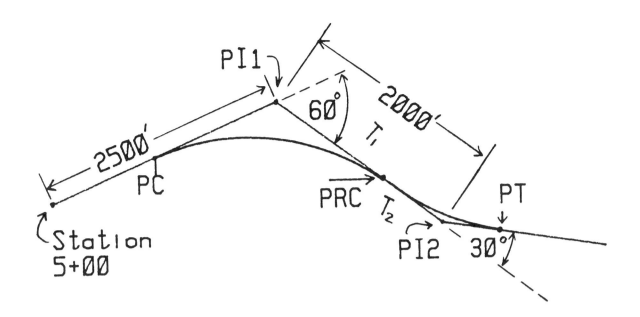

Problem E

17. $2000 = R (\tan 30° + \tan 15°)$; $R = 2366.025'$

18. $T_1 = 2366.025 (\tan 30°) = 1366.025'$

    $T_2 = 2366.025 (\tan 15°) = 633.975'$

    $(5+00) + (25+00) - (13+66.025) = 16+33.98 (=PC)$

19. $L_1 = R \Delta_1 = 2366.025 (60°\pi/180°) = 2477.696'$

    $L_2 = R \Delta_2 = 2366.025 (30°\pi/180°) = 1238.848'$

    $(16+33.98) + (24+77.696) = 41+11.68 (=PRC)$

20. $(41+11.68) + (12+38.848) = 53+50.53 (=PT)$

21. $(5+00) + (25+00) + (20+00) = 50+00 (=PI2)$

Problem F

22. $1000 = 2R \sin (60°/2)$; $R = 1000'$

    $T = 1000 \tan (60°/2) = 577.35'$

    $T' = 577.35 - 20 = 557.35'$

    $557.35' = R' \tan (60°/2)$; $R' = 965.36'$

23. $\dfrac{60°}{2} \times \dfrac{1}{2} = 15°$

Sample Problems 2

Problem A

1.
$$\frac{-0.4 - (0.8)}{-0.2} = 6 \quad ; \quad L = 600'$$

2. $820. - (3 \times 0.8) = 817.6'$

3. Elevation of the EVC: $820. - (3 \times 0.4) = 818.8'$

$$\frac{(818.8 + 817.6)}{2} = 818.2'$$

4.
$$\frac{(818.2 + 820.0)}{2} = 819.1'$$

5. $X = -g_1/r = 0.8/0.2 = 4$
$(18+00) - (3+00) + (4+00) = 19+00$

6.
$$Y = 817.6 + 0.8(4) + \frac{(-0.2) \times 4^2}{2} = 819.2'$$

7.
$$e = \frac{g_1 - g_2}{8} L = \frac{0.8 - (-0.4)}{8} 6 = 0.9'$$

or $820 - 819.1 = 0.9'$

8. $0.9 \times (2/3)^2 = 0.4'$

Problem B

9. $820. - (3 \times 1.25) = 816.25'$

10.
$$r = \frac{-2.75 - 1.25}{6} = -0.667 \, \%$$

11. $X = 1.25 / 0.667 = 1.875$
$(8+00) - (3+00) + (1+87.5) = 6+87.5$

12.
$$Y = 816.25 + 1.25 \times (1.875) + \frac{-0.667 (1.875)^2}{2} = 817.42'$$

Problem A

Using point A as the origin (0,0,0), and T = Breast of tunnel.
Calculate the coordinates as follows:

L.	Azimuth	Slope (%)	Dista. (ft)	Point	X	Y	Z
AB	300°	+1	300.	A	0.00	0.00	0.00
				B	-259.81	150.00	3.00
BT	30°	+2	100.	T	-209.81	236.60	5.00
AC	30°	+2	200.	C	100.00	173.21	4.00
CD	330°	+25	300.	D	-50.00	433.02	79.00

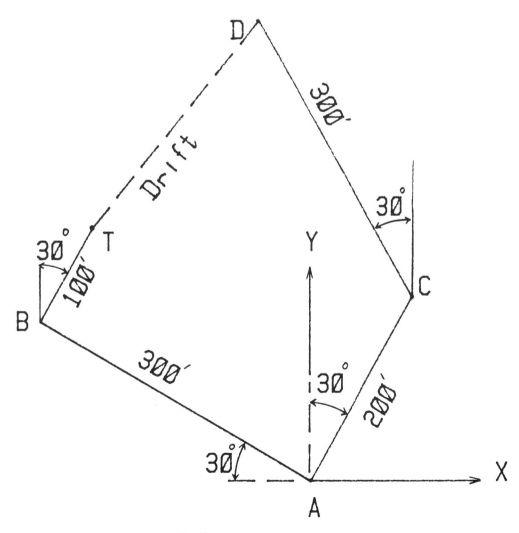

A-6

1. Azimuth of the drift = $\tan^{-1} \dfrac{209.81 - 50}{433.02 - 236.6}$ = $\boxed{39^{\circ}\,08'}$

2. Distance $\overline{TD} = \sqrt{(209.81 - 50)^2 + (433.02 - 236.6)^2}$
   = 253.22' (level)

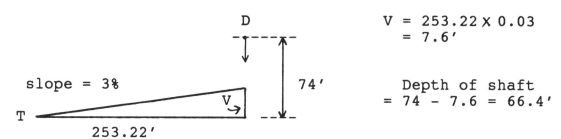

$V = 253.22 \times 0.03$
$ = 7.6'$

slope = 3%

74'

Depth of shaft
= 74 - 7.6 = 66.4'

T   253.22'

3. Slope length = $\sqrt{7.6^2 + 253.22^2}$ = 253.33'

4.   (1025N, 1575W) A

600

600   B (425N, 975W)

Problem B

5. $\angle$ A = 180° - (37°04' + 57°17') = 85°39'

7. Traverse Adjustment (By Compass Rule)

Line	Bearing	Distance	Latitude (cos)	Departure (sin)
AB	N 37°04'E	514.1	410.2	309.9
BC	S 68°32'E	1395.6	-510.7	1298.8
CD	S 55°40'W	961.3	-542.2	-793.8
DE	N 32°48'W	243.3	204.5	-131.8
EA	N 57°17'W	816.5	441.3	-687.0
Total		3930.8	+3.1	-3.9

Linear error of closure = $\sqrt{3.1^2 + 3.9^2}$ = 5.0 ft

8.   Precision = $\dfrac{5.0}{3930.8}$ = $\dfrac{1}{786}$ = $\dfrac{1}{800}$

A-7

10. See page 1-1 for radius point.

11. Double vernier:
    clockwise angle : 81°20′ + 07′ = 81°27′
      (counter-clockwise : 278°20′ + 13′ = 278°33′)

    Direct vernier : 21°10′ + 3′20″ = 21°13′20″

Sample Problems 4

1. Trapezoidal rule: Area ⟹ A

$$A = 10 \times \left(\frac{16.1}{2} + 12.9 + 23.2 + 24.3 + 23.5 + 20.8 + \frac{19.1}{2} \right)$$

2. Simpson's one-third rule: Area ⟹ A

$$A = \frac{10}{3} \times \left[16.1 + 2 \times (23.2 + 23.5) + 4 \times (12.9 + 24.3 + 20.8) + 19.1 \right]$$

3. Coordinates method: Area = $\frac{1}{2} \left(\frac{1}{43560} \right) \times$

$$\begin{aligned}
\Big[&- (2,250,000)(164,123.475) - (2,249,585.478)(164,382.551) \\
&- (2,249,350.149)(164,500) + (164,500)(2,249,585.478) \\
&+ (164,123.475)(2,249,350.149) + (164,382.551)(2,250,000) \Big]
\end{aligned}$$

Sample Problems 7

11. 94.8 − 4.1 + 2.1 = 92.8

12. HP = 82.48 + 2.0 + 20.0 + 10.0 + 22.0 = 136.48

13. CP#2 = 66.91 + 1.5 + 1.0 = 69.41

    Lt.HP = 66.91 + 1.5 + 11.0 = 79.41

14. Stake = 92.0 − 10.8 = 81.2

    F.L. = 81.2 − 1.0 = 80.2

15. 1091.17 − (4 + 8) × 0.02 = 1090.93
    1090.93 − (4 × 0.05) − (8 × 0.02) = 1090.57

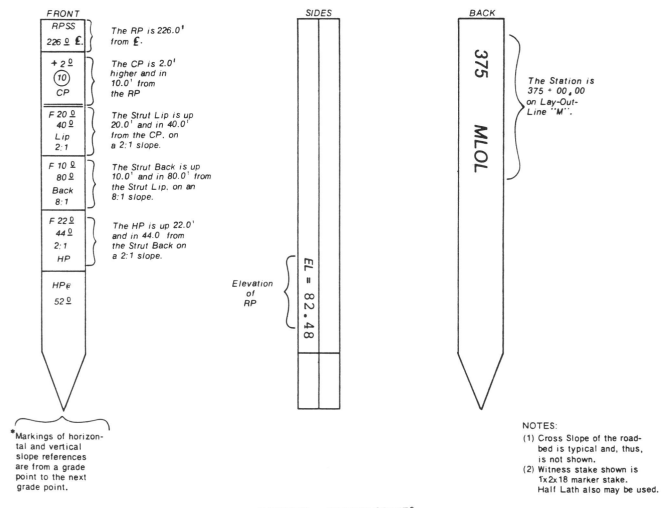

GEOMETRIC CROSS SECTION

MARKINGS on WITNESS STAKE*

*Markings of horizontal and vertical slope references are from a grade point to the next grade point.

NOTES:
(1) Cross Slope of the roadbed is typical and, thus, is not shown.
(2) Witness stake shown is 1x2x18 marker stake. Half Lath also may be used.

Fig. 11-05-G   Slope Staking - Strut Fill.

A-9

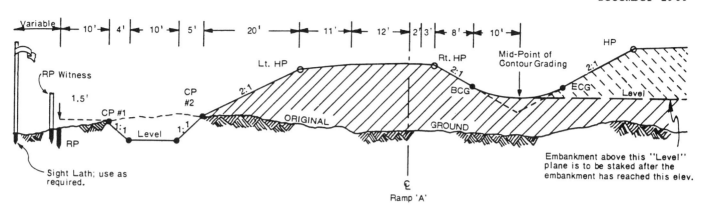

GEOMETRIC CROSS SECTION

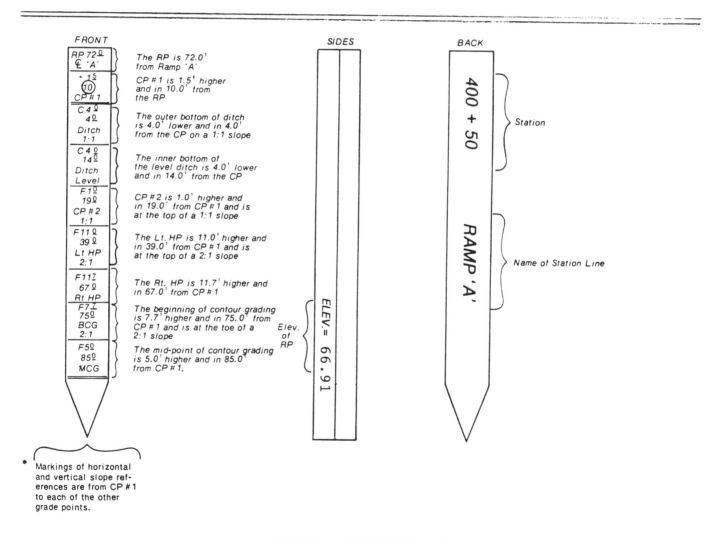

MARKINGS on RP WITNESS STAKE*

Fig. 11-05-E   Slope Staking - Split Roadbeds.

A-10

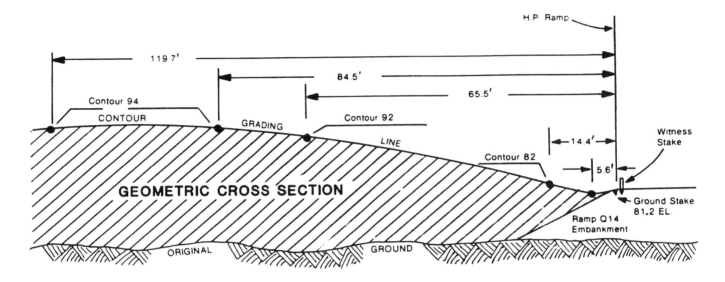

OFFSET GRADING STAKE @ Sta. 33~ , RAMP Q14

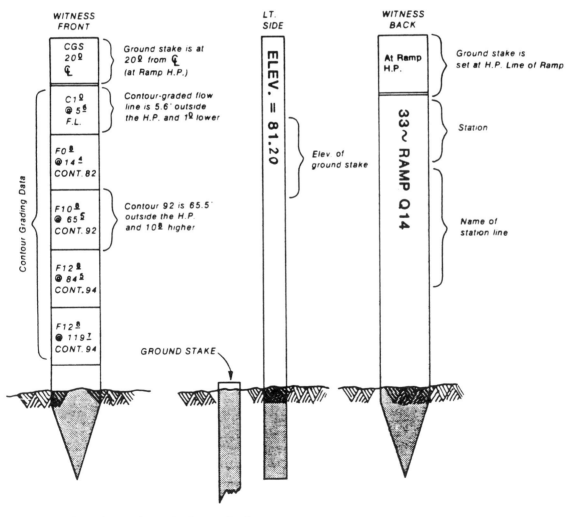

Fig. 11-08-C Offset Grading Stakes for Contour Grading

Sample Problems 8

1.

Station	Sewer	Alig.A	Exca.(-)	Alig.B	Exca.(-)
0+00	322.0	338.1	-16.1	356.3	-34.3
1+00	324.0	336.9	-12.9	348.2	-24.2
2+00	326.0	349.2	-23.2	339.3	-13.3
3+00	328.0	352.3	-24.3	326.7	+1.3
4+00	330.0	353.5	-23.5	334.8	-4.8
5+00	332.0	352.8	-20.8	351.3	-19.3
6+00	334.0	353.1	-19.1	356.1	-22.1

Excavation of Alignment A : by Trapezoidal rule,

$$V = 100 \times \left(\frac{16.1}{2} + 12.9 + 23.2 + 24.3 + 23.5 + 20.8 + \frac{19.1}{2} \right) = 12230.$$

Excavation of Alignment B : by Trapezoidal rule, (set +1.3 $\cong$ 0)

$$V = 100 \times \left(\frac{34.3}{2} + 24.2 + 13.3 + 0 + 4.8 + 19.3 + \frac{22.1}{2} \right) = 8980.$$

2. (1): $-0.26 \times \left[(4+25) - (1+45) \right] = -0.26 \times 2.8 = -0.728$
    (2): $+1.35 \times \left[(10+35) - (4+25) \right] = +1.35 \times 6.1 = +8.235$
    (3): $-10.5 \times \left[(12+50) - (10+35) \right] = -10.5 \times 2.15 = -22.575$

    Elev. = 348.26 - 0.728 + 8.235 - 22.575 = 333.19

4. BM2 = 32.45 + (BS sum) - (FS sum)
        BS sum = 3.27 + 4.56 + 5.34 = 13.17
        FS sum = 4.31 + 4.28 + 5.97 = 14.56
    BM2 = 32.45 + 13.17 - 14.56 = 31.06

    Difference in elevation between BM2 and BM3:
        BS sum = 6.72 + 4.58 + 8.21 = 19.51
        FS sum = 3.25 + 4.31 + 5.61 = 13.17
    (BS sum) - (FS sum) = 19.51 - 13.17 = 6.34

6.

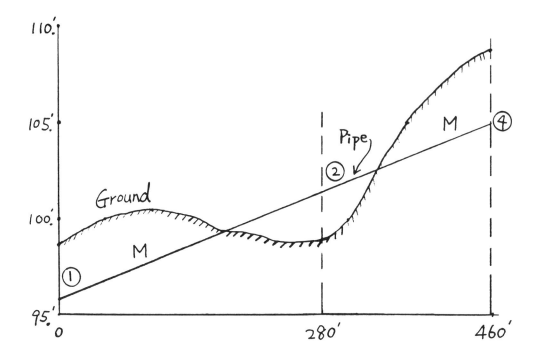

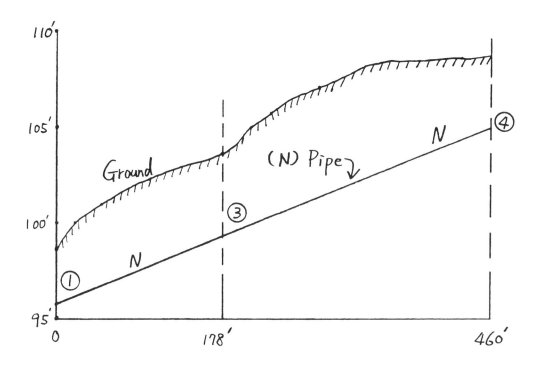

A-13

INDEX

✂

✂

Engineering Press Bookstore
P.O. Box 200129
Austin, TX 78720-0129

Engineering Press Bookstore
P.O. Box 200129
Austin, TX 78720-0129

Engineering Press Bookstore
P.O. Box 200129
Austin, TX 78720-0129